Bruce J. West

**Nature's Patterns and the Fractional Calculus**

# Fractional Calculus in Applied Sciences and Engineering



# Volume 2

Bruce J. West

# Nature's Patterns and the Fractional Calculus

DE GRUYTER

**Mathematics Subject Classification 2010**
Primary: 00A09; Secondary: 26A33, 35R11

**Author**
Dr. Bruce J. West
US Army Research Office
PO Box 12211
Research Triangle Park
NC 27709-221
USA

ISBN 978-3-11-053411-5
e-ISBN (PDF) 978-3-11-053513-6
e-ISBN (EPUB) 978-3-11-053427-6
Set-ISBN 978-3-11-053514-3
ISSN 2509-7210

**Library of Congress Cataloging-in-Publication Data**
A CIP catalog record for this book has been applied for at the Library of Congress.

**Bibliographic information published by the Deutsche Nationalbibliothek**
The Deutsche Nationalbibliothek lists this publication in the Deutsche Nationalbibliografie; detailed bibliographic data are available on the Internet at http://dnb.dnb.de.


Typesetting: VTeX UAB, Lithuania
Printing and binding: CPI books GmbH, Leck
♾ Printed on acid-free paper
Printed in Germany

www.degruyter.com

# Foreword

I have tremendous faith and hope in science and I try to study its evolutionary development over time, eras, and paradigm shifts. Science is not perfect, nor is it completely objective. It is, however, different than the many faith- and intuition-based elements of modern society. Science is not a philosophy that you get to pick and choose from a basket. Science is the light that can lead humans out of the darkness of our ignorance. Science adds knowledge and wisdom to the collective of human cognition to provide the world hope for the future. Popular books on science, like this one, add to that collective of scientific knowledge. Technical books in science, also like this one, light our path to a brighter future. West has performed both these roles in this book, Nature's Patterns.

Science, based on its underlying language and foundation called mathematics, evolves in its roles and importance to society. This steady evolution has brought society to a crossroads in its approaches to learning, thinking and decision making. We live in the post industrial age where information and systems science needs and advances drive progress in society. It used to be just knowing how to use a calculator was enough technical know-how to be considered a smart person. The complexity of modern life with its global competition and advanced warfighting technologies can threaten the health and wellbeing of all people and our entire planet. The main threat to progress is that humans, despite their incredible potential for thought and innovation, are often stuck in the morass of their pre-science past. People sometimes rely on instincts or beliefs, instead of trusting modern science with its reasoned evidence. Other people chose image-based influence marketing over the sometimes harsher or scarier reality of science. Mistrust and misunderstanding of science and its methods is a threat to future progress even though humans have benefited from science in many ways to build our current technologically-based and convenience-producing society. Society, through its education systems, needs to do a better job of advancing the appreciation, use, and development of scientific evidence and results. Considerations from the social and human domains must enter the sciences. This is where Bruce West's models and writings contribute.

Bruce West has the great knack of setting the stage with a mixture of philosophy and history. This approach helps to orient the reader towards a starting point and an ultimate goal. The start of this book (Chapter 1: What are we talking about?) doesn't disappoint in this regard. Physicist West immediately sets the stage for his quest – "Find an equivalent to force in the non-physical domain that is not mechanical, but no less real." He goes on to detail that his non-physical domains of interest will be the social and life sciences and that he ultimately hopes to quantify such forces. So right away, the reader sees that West is serious about this goal. He isn't going to just use the all-too-common metaphoric approach. He sees this new force as robust in its own information domain, not just a repurposed version of an existing physical force.

DOI 10.1515/9783110535136-201

Then he explains how complexity will help define this information force, but again he is careful not to think of his work as the usual form of trickle-down physics. I prefer West's goal for science "to garner understanding" than his alternate goal "to enable predictions", but it is his careful description of the role of mathematics in developing new modern science that excites me.

West bemoans that all too often in our quest to understand complex information that the limitations of mathematics compel the modeler to restrict the complexity level of the phenomenon being studied. I agree with West that mathematics does more than just build and solve the model, it also provides the language and logic to think about scientific phenomena. This need for more and better mathematical language is important in the development of all the sciences, including physics, social science, life science, and information science, and in the important role of decision making. West's work is as much mathematical theory as it is scientific.

The most compelling element of West's motivational and foundational chapter is found in his historical background in Natural Philosophy and Calculus. From that analysis comes his fundamental plan to reexamine, analyze, and redefine the fundamental concepts of space and time. By lifting the restrictions of inappropriately physical-based definitions of these fundamental ideas, mathematics and science can more appropriately contribute to the worlds of information, social, and life sciences. Connected to that effort is West's investigation to connect size and complexity, which in turn forms his science of allometry relations and the concept of functionality. From functionality comes the process of human decision making that demands more complexity and more use of nonlinearity in its models. West explains his conceptions and results both with underlying principles and examples. This combination of theory and application is a constant theme throughout the book.

West's models and measures of complexity are multi-scaled and nonlinear in their nature, often resulting in geometrical, statistical, and dynamical fractals. In West's framework, fractals and nonlinearities can appear in the modeling and analysis of both space and time. It is the local scaling of these fractal elements that help construct his measure of complexity. The long-held idea that analytic functions are the baseline for models and theories is no longer viable. West describes his concept of complexity as having "a strong component of regularity and an equally strong component of randomness." Therefore, complexity can be increased in a model or system by increasing degrees of freedom or by adding nonlinearities in its processes. Because we see and find extreme variability nearly everywhere, West concludes that we live in a complex world. Of course, that has been the case for a long time, but it is just now that we are embracing that complexity in our modeling and problem solving.

Information science affects the social and biological domains by modeling and analyzing how individuals, cells, organs, and organization interact, exchange ideas, share information, and conduct business. Society and organizations like the US military are attempting to use information processing and data analytics to make better and faster decisions. Researchers are making strides in their ability to model challenging problems and to understand the complex issues in our information-centric world. There is still much more to do in all modes of research – mathematical theories, modeling tools, scientific applications, and technical implementations. The farther in the future one looks, the more likely it is that information science dominates the research and commerce spotlight. A striking example of a significant information-science project is IBM's development of the Watson system to compete against humans in the information-centric TV game contest Jeopardy and subsequently used to aid in medical diagnosis and other large data analytic efforts. Watson's notable success illustrates the potential of information science to improve automated decision making systems.

After myriad cyber attacks and system hacks, the US government now recognizes its information systems and its communication networks as key parts of its critical infrastructure. Even in this era of systems such as Watson performing roles as information scientists and decision makers, there is still much more fundamental machine and human learning to be performed. Despite all this progress, significant deficiencies remain in the theoretical underpinnings of information science. The application of traditional reductive, linear techniques often results in sub-optimal results and limited tools. West has proposed the development of new, modern fractional mathematics to overcome some of these limitations.

The steady transition from physical to information science has impacted scientific thinking and methodology, teaching and learning, structures and processes, modeling and analysis. Human-related problems are often closely connected to information systems that behave differently than the physical properties and processes that were studied during the Enlightenment and Industrial Age. Complexity no longer reveals itself solely through the physical science-based lens of reductionism, higher-order derivatives, more dimensions, more accurate geometry, special functions, equations, and inequalities. Today, with the new information paradigms, we see complexity through a more human-focused, information-based lens with nonreductive measures for highly complex properties such as trust, behavior, cooperation, value, and influence. To continue this evolution in information science, West provides a more integrated, nonreductive, nonlinear framework and a set of Pareto-pdf-based tools to aid in modeling and decision making.

The application that West chooses to show the utility of his framework is allometry relations, the scaling relation between the size or functionality of an organ and the size of the entire organism. Through the extension of this relationship to information networks and subnetworks and the flow of information between the subnetworks, West's framework shows that information flows through the complexity gradient between the subnetworks. This Principle of Complexity Management (PCM) produces the information force sought by many information scientists. West's development and proof of this PCM uses fractional calculus.

The bulk of this book contains allometry examples that follow West's framework (the PCM), shows results achieved using the method, introduces models of allometry relations in various sets of data and applications, and the discusses the tools and theories available through fractional mathematics (calculus, statistics, and probability). West has given us a starting point to build mathematics and develop theories and measures for complexity, network processes, and information science.

USMA
West Point, NY
January 2017

Chris Arney

# Acknowledgement

Thanks to Adam Svenkeson for reading an early draft of the book and making valuable suggestions for its improvement. I also wish to acknowledge the love and encouragement of my wife Sharon and to dedicate this book to my four grandchildren: Alexandra, John Galt, Gabriel and Sydney.

DOI 10.1515/9783110535136-202

# Contents

# 1 Complexity

It is part of the mythology of book writing that the author should begin with a captivating first paragraph designed to capture the readers' attention and hold them transfixed. The myth is probably true for a work of fiction, but I am not convinced that it applies to an essay in science. On the other hand, there is no reason, other than the limited talent of the author, as to why a work of non-fiction cannot be written as interestingly as one of fiction. So I begin in the same direct manner I will later present the science, with the hope that what it lacks in literary craftsmanship, it makes up for in clarity and in providing a glimpse into the excitement of scientific discovery.

Scientists measure things, including the positions of stars, the amount of rainfall and the heartrates of patients. The numbers collected quantify the phenomena the investigator wants to understand. An astronomer deduces whether the faint dot overhead is a fixed star, a moving planet or a rocketing comet, using the physical theories of celestial mechanics. The numbers tell the meteorologist if there is a pattern of increasing or decreasing rainfall and whether that pattern indicates an organized change in the weather or is the result of chance. A physician determines whether the pattern of hearbeats reveals if the patient has a cardiovascular problem that requires intervention, or if s/he is having an anxiety attack. Each science organizes measurements in ways that communicate the most to the practitioners of that particular discipline. These disciplines taken as a group constitute the scientific view of the world, which is to say that if it is a matter of science it ought to be quantifiable.

But given the frequently sharp boundaries between scientific disciplines are there quantifiable patterns that appear in multiple disciplines, but which are typically discussed without reference to, or acknowledgment of, one another? For example, are there empirical relations in biology that are identical in form to heuristic patterns found in urban dynamics, or are parallel to those revealed in psychology, or excavated in geophysics? The short answer to this question is an unqualified yes. These phenomenological relations and others, taken as a group, are the topic of this essay. I am concerned with what these patterns are, how they are obtained, why they are formed, what they tell us about the phenomena under study, and how they can be related to a property that is common to each of the disciplines.

My thesis is that there is a class of natural patterns, tagged with the rubric *allometry relations*, which is a consequence of complexity, but is essentially independent of the disciplines in which they are found. That is not to say that there is not a discipline-specific mechanism that can explain a given allometry relation (AR), but that such a mechanism must have a certain level of complexity, and it is the complexity of the mechanism which is the actual origin of the pattern. So let us begin by developing an intuition of what constitutes complexity in today's world.

DOI 10.1515/9783110535136-001

## 1.1 Some perspective

It is difficult, if not impossible, to recapture the mind set held by a population prior to its adoption of scientific reasoning. Our industrial society is so intertwined with scientific/technological concepts that separating the social from the technical is extremely difficult to do. This is not just recognition of the technological artifacts that control our lives, such as: talking on cell phones, tracking down a fact on the Internet, driving a car, using air conditioning, playing computer games, the list is nearly endless. The point is, if a modern person were placed in a hostile environment, without their technological devices, could they survive, or would they be amazed by how most survival problems have been reduced in our society to a level of discomfort by technology and the problems' significance to an irritating abstraction! Technology actually determines how we see and understand our world, as well as, how we respond to it. The medium is no longer the message, with my apologizes to Marshall McLuhan, technology determines how we think.

This essay attempts to show that the reasoning and methodologies, which have historically been so successful in explaining phenomena in the physical sciences, have very restricted domains of success in the life and social sciences. We acknowledge at the outset that underlying the physical arguments is the idea of forces of various kinds; mechanical, electromagnetic, gravitational and so on. But these forces do not appear outside physical science and although there has been ongoing efforts to construct social and biological analogues to physical forces, the fruits of these efforts have primarily been restricted to the role of metaphor. We know that the mechanical forces that dominate the physical world, are not sufficiently robust to explain phenomena in the biological and social worlds. Therefore, it is necessary to find a force-equivalent that exists in the non-physical domain in which we live; a 'force' that is not mechanical, but is no less real.

So the question is: Are there real forces, with predictable outcomes, which operate in the social and life sciences. Forces that are not merely metaphors for those in the physical science? Going even further: If such forces exist, can they be quantified?

The answer to these and similar questions is a tentative yes.

In the following chapters we explore how the concept of entropy, with its relation to randomness, information, complexity, and the asymmetry of time, is used to develop the idea of an information force. Such a force is shown to be generated by the interaction between two complex systems, as a consequence of the relative complexity imbalance between them. The existence of an information force is almost a new scientific idea, which means that some investigators have identified its effects in various context, but failed to identify information gradients as the source of the effect. Consequently, we present extensive justification for the concept. My remarks are intended to provide historical and contemporary, even if preliminary, evidence for the existence of an information force.

With a view towards justifying the existence of an information force, we assert that science is undertaken in order to make the uncertain less surprising, and to enable prediction. Science is intended to garner understanding of the world, in everything from generating electricity for the power grid using sun, wind and water; to providing insight into the social world, from orderly elections in some nations to the overthrow of the government in others; to penetrating the mysteries of medicine, from cancer to loss of equilibrium; and on and on. Major successes have been achieved, using the predictive methods of science, over the past few hundred years. But the successes have been circumscribed by the quality of the models with which the phenomena have been described, which in turn have been limited by the robustness of the underlying mathematics. Thus, the mathematics used often determines the complexity level of the phenomena that can be faithfully modeled.

Few people know that the phrase 'trickle-down economics' originated with the depression era humorist and political satirist, Will Rogers, whose statue stands in the corridor, between the rotunda and the House of Representatives, where he continues to oversee those that were the target of his humor. This 'trickle-down theory' of the redistribution of wealth suggests that tax breaks, or other economic benefits, provided to businesses and upper income levels, will benefit the less fortunate members of society by improving the economy as a whole. The source was a quip that Rogers made with his characteristic sarcasm:

> The money was all appropriated for the top in hopes that it would trickle down to the needy.

This specific observation regarding the transfer of 'wealth' was broadened by Latka et al. [181], beyond economic wealth to scientific wealth, asserting that most physicists are proponents of "trickle-down physics". They suggest that physicists believe that the development of medicine and social science can be, and often is, determined by adopting the methods and techniques developed in the physical sciences. In fact, it is difficult to envision modern medicine without sophisticated imaging, such as computer tomography, nuclear magnetic resonance imaging and radiography that extends the life of thousands of cancer patients every year. Similar contributions to other disciplines can also be identified.

Physicists often find themselves at the frontiers of medicine, sociology, and other disciplines, outside, but adjacent to the physical sciences. It is often not clear whether being at such interfaces is intentional or not. But regardless of how they got there, such interdisciplinary research involving physicists is reciprocal, so that physics benefits from the interaction, as well as, the non-physical discipline. As they acknowledge, the paper by Latka et al. [181], is a testimony to that reciprocal benefit.

However, the mathematics that physicists employ, does more than simply provide models of the phenomena they seek to understand. Mathematics provides the elements of language and the logical tools necessary to think about those things in a way that allows them to make new, interesting and testable predictions. Consequently, the complexity of the problems investigators desire to solve, challenge existing mathematics and often specifies the tools that need to be developed. This is one source for the criteria of the new mathematics that is needed. But, of course, mathematics is itself a science and mathematicians create new areas of study to satisfy their own curiosity, without any particular disciplinary application in mind. For example, when Euler (1707–1783) invented graph theory, he was not thinking of social applications, he was interested in answering a mundane, but intellectually challenging, question about the order in which the bridges in a small German town could be crossed by following the shortest path.

*Natural Philosophy* was the intellectual forerunner of science and its intent, since the earliest recordings of the Greeks, had been to develop wisdom. The introduction of mathematics into Natural Philosophy by Newton [232], the action that transformed it into science, changed its purpose, from the pursuit of wisdom, to the acquisition of knowledge. The area of mathematics that did more than any other to facilitate this transformation was the differential calculus, with its ability to answer questions about the earth and the cosmos, with equal agility. Answers that could be experimentally tested and either falsified or verified.

## 1.2 It started with physics!

The calculus was invented by Sir Isaac Newton (1642–1727) during the years of 1664–1665. At that time, Cambridge University, where he was a student, was closed to reduce the devastation of the Great Plague and Newton sequestered himself at home, conducting the research that would eventually establish him as the world's leading scientist. During this period of self-imposed isolation he, then 22 years old, laid the foundations for both the differential and integral calculus, which he called the method of *fluxions* and the inverse method of fluxions, respectively. This new mathematics was detailed in *The Method of Fluxions and Infinite Series*, which although completed in 1671, was published posthumously in 1736 in a translation from the Latin by John Colson (1680–1760), due to Newton's unwillingness to publish any of his results. Colson was Newton's successor as the third Lucasian Professor of Mathematics at Cambridge University.

Newton's reluctance to publish fueled, in his later years, an acrimonious priority dispute with Gottfried Leibniz (1646–1716) over who first discovered (invented) the new mathematical language. The debate was carried out mostly, but not exclusively, through surrogates. Looking back over the more than 300 years, it is probably more important to note the fragility of scientific egos than it is to provide either of them with additional posthumous accolades. From today's perspective it is probably safe to say that the two men invented the calculus mostly independent of one another, but the elegance and simplicity of the Leibniz notation resulted in its general adoption by the scientific communities and is still used today.

It is remarkable that the Marquis de l'Hôpital recognized that Newton's *Principia* [232] was "a book dense with the theory and application of the infinitesimal calculus"; an observation also made in modern times by Whiteside [375]. At the time of its publication the content of Newton's *maximus opus* was understood by only a handful of mathematicians, including Leibniz, Huygens, Bernoulli and de l'Hôpital. However, it must be noted that the underlying concepts of the fluxion-based infinitesimals were presented, not to say disguised, as traditional and often impenetrable geometrical arguments that relied heavily on geometrical limit-increments of variable line segments.

So why dredge up this history about the differential calculus? The reason is the success that Newtonian mechanics has enjoyed in describing the physical world. The calculus that Leibniz created would have been an extraordinary intellectual achievement independently of any potential applications. However, the assumptions made regarding the functions involved in the development of the mathematics would not necessarily be of particular interest to a physical scientist. On the other hand, the fluxions of Newton were constructed to provide a theory of motion and to define and understand the forces of nature, so the underlying assumptions regarding the functions used, relate directly to the empirical properties of the real world. Recall that there was no clear definition of a mechanical force prior to Newton's introduction of it in the context of motion of physical (celestial) bodies.

In the preface to the first edition of the *Mathematical Principles of Natural Philosophy* [232], published in 1686, Newton reveals his purpose:

> ...I consider philosophy rather than arts and write not concerning manual but natural powers, and consider chiefly those things which relate to gravity, levity, elastic force, the resistance of fluids, and the like forces, whether attractive or impulsive; and therefore I offer this work as the mathematical principles of philosophy, for the whole burden of philosophy seems to consist of this – from the phenomena of motions to investigate the forces of nature, and then from these forces to demonstrate the other phenomena...

*Book 1* of the *Mathematical Principles of Natural Philosophy* begins with a number of definitions; two of which changed how the science of mechanics and in fact all of physical science were to be subsequently understood. Immediately following the section on definitions, Newton introduced his three Laws of Motion as axioms: Law I follows from his definition of inertia and Law II follows from the definition of force, but nowhere in the three books do we find the explicit equation for the force $\mathbf{F}$ in terms of the mass $m$ and acceleration $\mathbf{a}$: what is now in every freshman physics text, $\mathbf{F} = m\mathbf{a}$. Following his eight definitions Newton interjected what he called a marginal note, or *scholium*, noting that he had defined quantities, such as force, that were not well known, but had refrained from defining the all too familiar quantities of space and time.

In his scholium, he clarifies that the motion with which he is concerned involves mathematical time that is not related to anything in the physical world, but flows continuously, without distortion or interruption. This true time is distinct from the sensible time that is broken up into seconds, hours, days and years. In the same way he goes on to describe absolute space, which is independent of anything external and is immovable and which he distinguishes from relative space. His idea of space is where objects are contained, but in itself is featureless.

Much of the discussion in the present book is a consequence of replacing these, now atavistic, definitions of space and time. They are outdated, in part, because they fail to capture the rich structure of the complex phenomena of interest in the modern world. The failure to go back and reexamine these fundamental assumptions have, in large part, restricted the extension of the modeling techniques of physics into the non-physical worlds of psychology, sociology and the life sciences. The experience of space and time differs between that of the claustrophobic and the agoraphobic, from the performer on the stage to the surgeon operating on the brain, from the warrior on the battlefield to the physician on the critical care ward. We require a mathematics that can capture all of this and much more.

## 1.3 Complexity and mathematics

The physical world of Newton was relatively simple; by which I mean simple in its mathematical description, but simultaneously profound in its capacity to explain observations and make predictions. Over the course of time the instruments developed to test these predictions have become so refined that we now directly experience the effects of things whose fundamental nature we neither know nor understand. But we have every confidence that there exists a substantial collection of people who have made it their life's work to understand such things. What is becoming increasingly clear to this group is that the technical intensity of the world has become so dense that the mathematical language initiated by Newton is no longer adequate for its understanding. In fact we now find that we have been looking at the world through a lens that often suppresses the most important aspects of phenomena, most of which are not "simple". These are characteristics of the phenomena that cannot be described using differential equations and we refer to them as complex.

Most things that we label as complex have two dominant characteristics; they have a strong component of regularity and an equally strong component of randomness. Take, for example, the weather. If it was sunny and 80 °F yesterday, it will probably be the same today. But do not bet your life on it. It could also rain. Consequently, complex phenomena, such as the weather, can be defined in terms of a balance between regularity and randomness [347]. On the other hand, the stability and adaptability of a complex process can be lost through an imbalance favoring either one or the other. Therefore, extending the modeling of physical phenomena to the behavior of living organisms, either individually or collectively, as done by scientists in the nineteenth century, was almost uniformly disappointing. The initial successes of physics modeling relied in large part on Newton's concept of force, which has no direct cor-

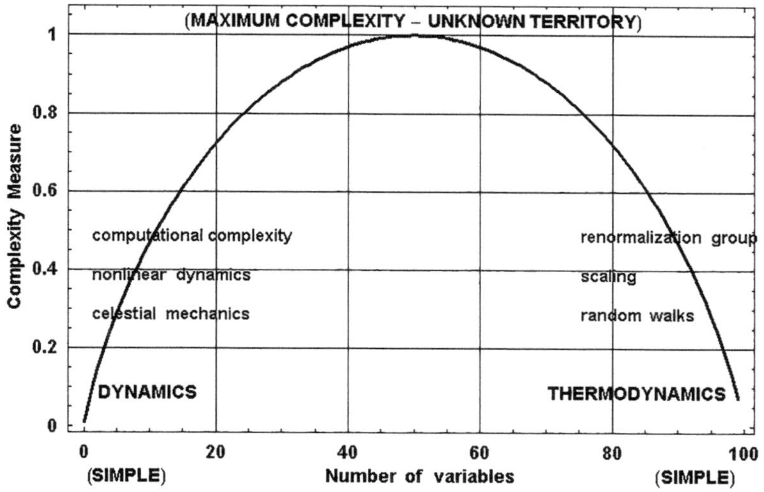


**Figure 1.1:** Here is a sketch of a possible measure of complexity. At the left where there are few variables determinism and the dynamics of individual trajectories dominate. At the right where there are a great many variables, randomness and the dynamics of probability densities dominate. In the center, at the peak of the complexity curve, maximum complexity involving both regularity and randomness together, with neither dominating. (From [362] with permission.)

respondent in social sciences, except in a metaphorical sense, and has only selective utility in the life sciences. Consequently, the genesis of non-physical forces, outside the physical sciences, must be traced to non-physical sources.

Complexity is implicit in the system-of-systems or network-of-networks concept. One way to talk about complexity that has been repeatedly used is given in Figure 1.1 [362]. This figure introduces a measure of complexity that starts at zero, increases to a maximum value as the number of elementary or micro-variables describing the system increases, passes through a maximum, and then decreases to zero again. Start at the left of the figure, with the dynamics of one or a few variables and denote this as being simple, since the equations of motion and their solutions are well known. Complexity increases with an increasing number of variables, or degrees of freedom, as we proceed up the curve going from left to right. The mathematics of such systems includes, mechanical forces, nonlinear dynamics, control theory and so on, referring to mathematical formalisms that are fairly well understood. There has been a substantial body of mathematical analysis developed regarding complexity and its measures, and the broad range over which mathematical reasoning and modeling have been applied is rather surprising. One class of problems, which defines the limits of applicability of such reasoning, is "computational complexity" the distal discipline on the left side of the complexity curve. A problem is said to be computationally complex if to compute the solution one has to write a very long algorithm, essentially one as long as the solution itself. Applications of this quite formal theory can be found in a variety of areas of applied mathematics, but herein I avoid these more formal issues and focus attention on the influence of nonlinear dynamics in such theories and the subsequent notion of complexity implied by this influence.

The common feature of all the techniques on the left side of the peak of the complexity curve is they are primarily methods for handling the deterministic single particle trajectories of the system's microscopic dynamics. However, as the complexity continues to increase with an increasing number of micro-variables, the techniques become less useful and blend into what we do not yet know how to mathematically manipulate. What becomes evident is that reversible determinism alone is no longer sufficient to understand truly complex phenomena.

On the other side of Figure 1.1, where the number of microscopic degrees of freedom is very large (infinite), we have equilibrium thermodynamics, and the phenomena are again simple. The mathematics describing such systems involve partial differential equations for the evolution of the probability distribution functions (PDFs) in term of macro-variables, renormalization group theory (RGT) and scaling of the coupling across scales, all of which are mathematical formalisms that assist our understanding of the physical and life sciences. The mathematics here is designed to handle irreversible random phenomena and describe the system's many-body dynamics. Here again ascending the complexity curve, this time moving from right to left, the system being modeled increases in complexity with decreasing number of micro-variables, but an increasing number of macro-variables, and the mathematical tools available again become increasingly less useful. The symmetric form of the curve is notional, which is to say, we do not know its exact shape. However, the curve's non-monotonic behavior captures general properties of complex systems, which are known.

The scientific territory of the unknown lies between these two extremes of simplicity, depicted in Figure 1.1. The area of maximum complexity is where one knows the least mathematically, but often there exists quite a bit of experimental data. The peak is where neither randomness, nor regularity, dominate; nonlinearity is everywhere; all interactions are nonlocal, and nothing is entirely forgotten. Here is where turbulence lurks, where the mysteries of neurophysiology have taken root, and the dynamic secrets of DNA remain hidden. All the problems in the sciences of the animate and inanimate, which have confounded the best minds for centuries, are here waiting in the dark corners for the next mathematical/scientific concepts to shed some light.

This is the picture of complexity I had in mind when the concept was introduced into our discussion. Complexity is a balance between regularity and randomness, with the dynamics having elements of both [364]. Notice that the concept of randomness was inserted without fanfare and is either as simple, or as complicated, as you choose to understand it. In any event randomness certainly warrants systematic discussion. So let us consider how uncertainty first entered the sciences through the introduction of statistics and probability into the theoretical understanding of physical complexity.

### 1.3.1 Machine age

At a time in our history when most of the world's population lived on farms, flying was restricted to birds and the earth's winds powered the international shipping lanes, the mathematicians, Gauss [111] in Germany and Adrian [1] in the United States, introduced into science the normal probability density function (PDF), which subsequently became known as the Law of Frequency of Errors (LFE). The LFE established that the variability found in the results of even the simplest experiments obey a natural law, which was to become the lens through which the distortions produced by complex phenomena were seen for over two hundred years. The year following the publications of the mathematical results, the physicist Laplace [177] proved the Central Limit Theorem, which was the first of many proofs that under a given set of conditions, data sets are described by the bell-shaped normal PDF.

Adrian and Gauss independently explained the mystery as to why physical experiments never give the same result twice. Experimental data is observed to cluster around a central value, but there is always variability, whose explanation had eluded investigators for centuries. The two mathematicians hypothesized that the arithmetic average was the best representation of the data and the bell-shaped curve, depicted in Figure 1.2, characterizes the variability around that average value. This introduction of statistics into experimental science revolutionized not only how scientists thought about the world, but how complexity and uncertainty were subsequently to be understood.

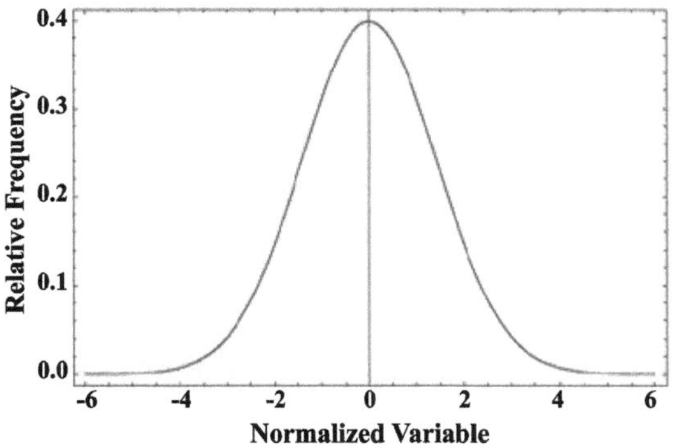


**Figure 1.2:** The bell-shaped curve is the normal distribution of Gauss and Adrian. The universal form shown is constructed from data by subtracting the average value, centering the resulting curve on zero, and dividing by the standard deviation. The numbers denote the distance from the average in terms of the number of standard deviations.

It is remarkable that the random variations in measurement are not capricious, but follow an unexpected law; the LFE. The name derives from the fact that the empirical laws of physics are all expressed in terms of averages. Consequently, the predicted behavior of a simple physical system is interpreted as being that of the average, the mode of the PDF. Therefore the average is the predicted value and the deviation of the measurement from the predicted value must be error. In this world view there exists a correct value for the outcome of an experiment and although accurate for simple physical systems, this view imposes an unrealizable constraint on how we understand the empirical world.

In any event, the success of the normal PDF in physics led to its adoption in explaining non-physical phenomena as well. At the intersection of physics and sociology the 'average man' was introduced, where physics met economics the concept of the 'rational man' was adopted, and the 'reasonable man' was the love child of physics and jurisprudence. The normal PDF captured and held captive the imagination of scientists throughout the world for over two centuries. The bell-shaped curve was accepted, in part, because it allowed practitioners of the various non-physical disciplines to insert uncertainty into their discussions in a manageable way. A little uncertainty was not only tolerable, but welcome. It provided a patina of reality to overlay modeling exercises.

These mathematical ideas found fertile ground in the assembly lines of the Machine Age. From the tolerance of a crank-shaft, to the quality control of the assembly-line output, the artifacts of the Machine Age lent themselves to the description of the normal curve. Two hundred years of gathering data has taught us the properties of systems described by the normal curve. First and foremost, such systems are linear, so a system's response is proportional to the strength of the stimulation. A 10% excitation produces a 10% response, or maybe a 20% response, but nothing too crazy. The system remains stable when excited, and therefore, in principle, its behavior can be predicted, at least in a fairly narrow, probabilistic sense. A probabilistic prediction means that a given fraction of a large number of identically prepared systems, each perturbed in the same way, will have the same proportionate response. Secondly, such systems are additive, they can be reduced to fundamental elements, which weakly interact and recombine, after being perturbed, to reconstitute the overall system. Consequently, linear additive phenomena that are stable when stimulated, describe the world of Adrian and Gauss [337].

This is the world view of the typical Westerner. The machines that populate our world are the result of the normal curve dominating manufacturing to reduce variability and we have come to expect that refrigerators, dishwashers, automobiles, and all other machines of a given make and model are identical. But not just machines, the clothes and shoes we wear, as well, have a similar uniformity. The variability is controllable and reducible to the point of vanishing by the process of manufacturing. This anticipated controllability seeps into the individual psyche to reinforce the normal curve. We are annoyed when airplanes are delayed, surprised and angered when a newly purchased appliance does not work properly, and outraged and frightened when our financial identity is stolen. The farther the event is from the expected value, the greater is our irrational (emotional) response. We neither anticipate nor accept when the technology of our world no longer works for us, much less when and how it can be turned against us.

The culture does a great deal to reinforce this idealized linear view of the world. Every student who has taken a freshman-level course in the sciences has been graded on a curve, which is to say, the grades were forced to conform to a normal curve, as depicted in Figure 1.2. Grading on a curve means that 68% of the students receive *C*s, another 27% are equally divided between *B*s and *D*s, and the final 5% share the *A*s and *F*s. After taking enough of these classes, students begin to accept the reality that something as complex as learning, particularly in the sciences, can be represented by a linear additive process, which of course it cannot. However, social pressure forces us to continue to act as if it were true, even though we know experientially that it is not.

Thus, Western culture, both directly and indirectly, has taught the same lesson to the janitor, as it has to the CEO of any large corporation, and all people in between. The repeated lesson is that there is always a best way to accomplish a task; a small amount of variability can be tolerated, but not too much; the lesson of the normal curve. This is how Westerners are, not so much trained, as they are seduced into a Machine Age way of thinking. Therefore, due to continual reinforcement, any important decision is only acceptable if it does not deviate very much from what a group of reasonable people would do. We are fortunate that historically such people as Edison, Salk and Einstein, did not consult such groups.

### 1.3.2 Information age

In the world of Adrian and Gauss, which developed into the Machine Age, the ideal emerged that everyone should make the same amount of money for the work they do. Some might make a little more, and others a little less, but the average salary ought to be a good indicator of the health of the economy and of equity within the society. In such a world everyone would be a mediocre artist; play a musical instrument badly; all sports would result in a tie and there would be no heroes. No one would be outstanding and everyone would be nearly equal and equally uninteresting. But that is not the world we live in. In our world there is Michael Jordan who destroyed his competition in basketball; Picasso, who turned the art world on its head; Miles Davis, who captured the listener's soul with his music, and people who make staggering amounts of money as income. In the real world there are stock market crashes and economic bubbles; earthquakes and brainquakes; peaceful demonstrations and uncontrolled riots; some children die and others live to be very old. There is extreme variability everywhere and in everything.

The point is that we live in a complex world, but our understanding and ways of thinking are based on simple models, more appropriate to the Machine Age. The distribution of income is not a bell-shaped curve, with a well-defined average and standard deviation. It is a curve that has a very long tail, as shown schematically in Figure 1.3 for positive values of the variable. This long-tailed distribution, an inverse power law (IPL), first discovered at the end of the nineteenth century by the engineer turned sociologist Vilfredo Pareto, is more appropriate for the Information Age. The Pareto PDF [240] captures the extreme variability that defines the most important events in our lives.

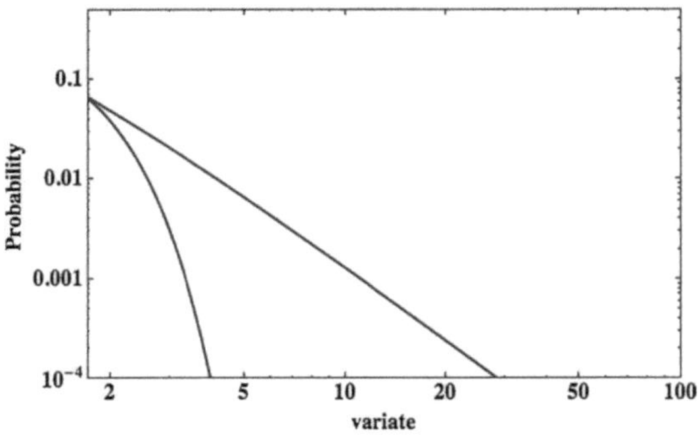


**Figure 1.3:** The bell-shaped curve of Adrian and Gauss is compared with the IPL of Pareto. On this log–log graph paper the normal curve is parabolic and the Pareto curve is a straight line with negative slope. The much longer tail of the Pareto distribution is evident, indicating that extreme events are much more probable in the latter case than in the former and consequently are much more important.

Without going into mind-numbing detail, it is clear that at the four standard deviation level of income (horizontal axis), depicted in Figure 1.3, the IPL PDF has at least two orders of magnitude greater probability of a person making that income (vertical axis), than predicted by the normal curve. The Pareto PDF corresponds to those in the upper few percent of the income distribution and the existence of this tail in the data persists into the twenty-first century in Western society. The Pareto IPL has a disproportionately small number of people amassing a disproportionately large fraction of the income. These are the super rich out in the tail of the distribution. The imbalance in the PDF was identified by Pareto to be a fundamental inequity, from which he concluded, after exhaustive analysis, that society was not fair. The Pareto world view, based on the IPL, is very different from that of Adrian and Gauss, based on the normal curve. This imbalance is a fundamental feature of the Information Age and is a manifestation of complexity, as we subsequently explain [363].

The complex phenomena of biological macro-evolution, letter writing, turn-taking in talking, urban growth, and making a fortune, are more similar to one another statistically and to the distribution of Pareto, than they are to the distribution of individual heights within a population. The chance of a child selected at random, from a large group in a Western city, getting rich is very small, but it is still much greater than the chance of that same child becoming very tall. A person's height is a hard ceiling determined by nature, whereas a person's wealth is a much softer upper limit that allows for the most dedicated to overcome the social barriers restricting income levels.

The Pareto PDF burst into the general scientific consciousness at the turn of this century, with the realization that it described the connectivity of individuals on the Internet and the World Wide Web. Then came a flood of insights that such IPL networks could model the failure of power grids, decision making, groupthink, brainquakes, turn-taking in conversations, habituation, walking, and a broad range of other phenomena in the social and life sciences. What caught the attention of many was the fact that such networks were robust against attacks for which they had been designed, but they were fragile when confronted with directed attacks, or with unanticipated challenges. A common feature of all these network models is their dependence on information and information exchange, rather than the energy exchange that dominated Machine Age thinking. The new Information Age dynamics entail a new way of thinking; new, because of the complexity-induced imbalance manifest in the IPL PDF.

A word of caution is in order before we continue. The IPL arises in a wide variety of contexts and although they all have the imbalance noted by Pareto, the causes for the imbalance are as varied as the phenomena in which they arise. There is one road to the normal curve of the Machine Age and it is straight and narrow, if somewhat steep. However, there are many roads to the Pareto PDF of the Information Age; some are wide and smooth, others are narrow and tortuously convoluted, but they all are the result of complexity in its various guises. Let us consider a couple of familiar examples to help clarify what we mean.

Scientists cannot predict the magnitude of the next earthquake, nor of the next brainquake [235], nor the time when they will occur, because of the extreme variability of such complex phenomena. If quakes were governed by the normal curve, its average size would be determined and could be predicted with some degree of certainty. In a related way, once a quake of a given size occurs, investigators would be able to predict how long they would have to wait before another of comparable size occurs. As depicted in Figure 1.4, the number of earthquakes of a given magnitude follows an IPL, named after the seismologists Gutenberg and Richter [134]. The same IPL structure was found by Osorio et al. [235] in their study of the size and frequency of epileptic seizures, as reviewed by West [360].

An increase by one unit on the horizontal scale corresponds to a factor of ten increase in quake magnitude. Such an increase is the difference between crawling by a hidden police car on a downtown street at 10 mph, versus roaring by him with your foot to the floor at 100 mph. On the other hand, the energy associated with a unit change denotes a factor of approximately 32, or the difference between jumping off a one story building, versus leaping from the roof of a thirty story building. A mouse might survive both falls, but a person would walk away from the first, but be splattered on the sidewalk in the second. A single unit on the appropriate scale can make a catastrophic difference.

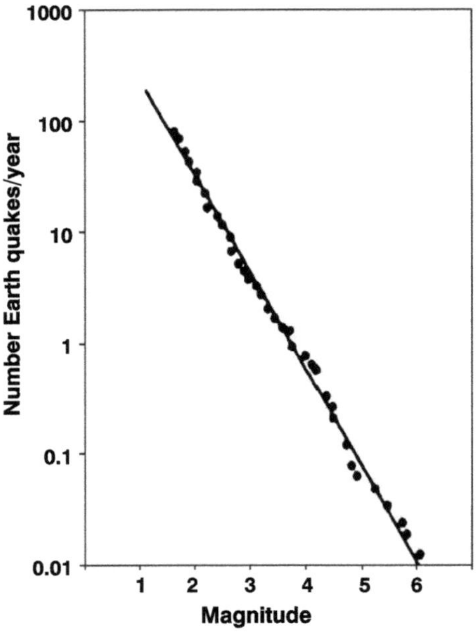


**Figure 1.4:** The estimated number of earthquakes in the world of a given magnitude since 1900, from the US Geological Survey National Earthquake Information Center. The solid line connecting the data points is the Gutenberg–Richter Law for the frequency of earthquakes of a given size with an IPL with a slope of −1.

How long one waits for an event of a given size to appear depends on the probability of such an event occurring. In fact the waiting time depends on the inverse of the probability of the occurrence of such an event; very probable events occur frequently and we see them quite often, whereas improbable events are rare and may only appear to us in history books. Consequently for a process described by Gauss statistics the probability of an event size far from the average is exponentially small and therefore we would expect to wait a very long time for such an extreme event. It is clear from Figure 1.3 that the probability of an extreme event is much greater for a Pareto than for a normal process and therefore extreme events occur much more frequently in the former than in the latter process.

It should also be stressed that the distribution of time intervals, between earthquakes of the same size, was empirically determined by Omori in 1894 to be an IPL with index −1 [234]. We do not display Omori's Law, but note that it has the form shown in Figures 1.3 and 1.4. The same recurrence distribution describes those of brainquakes [235]. Like the distribution of income, where those individuals at the high income end dominate the social situation, the extreme magnitude quakes dominate the geophysical and physiological situations. The damage done by the recent magnitude 8.9 earthquake off the east coast of Japan dwarfed the effects of the much more frequent magnitude 7 quakes. Note that a 2.9 magnitude quake is a million times more frequent than is an 8.9 magnitude quake, as well as being a million times weaker in magnitude.

Consequently, if city planners used incorrect statistics for earthquakes, the building codes designed to withstand extreme quakes, thought to occur every 200 years, might instead occur multiple times within the lifetime of a single individual. This would be a classic mistake; thinking about a complex phenomenon as if it were a simple one. But aside from getting the safety factors right, what else can we do to predict and prepare for natural disasters, like earthquakes; or for man-made disasters, like war, for that matter?

There is a great deal we can do in terms of anticipation and preparation for the effects of complex phenomena that are man-made. We can do this by modifying our thinking to adjust to the world of Pareto, that of the Information Age, and leave behind the counter productive world view of Adrian and Gauss, that of the Machine Age.

We are interested in how the models of complexity are transformed from those adopted using the world view of linear, small variations implied by the normal curve to that of the nonlinear, richly complex variations implied by the Pareto IPL PDF. To identify what needs to be done requires that we examine what is entailed by Pareto's identification of the imbalance in the underlying processes. This is particularly true given that the complexity in the social and life sciences are almost without exception described by IPL PDFs. We take up this challenge in Chapter 4 after reviewing its empirical aspects in Chapter 3.

### 1.3.3 Information force

In a recent paper, Piccinini et al. [248] observed, that in the middle of the last century, the mathematician Norbert Wiener [376] speculated that a system high in energy can be controlled by one that is low in energy. The necessary controlling force is produced by the low energy system being high in information content, and the high energy system being low in information content. Consequently, there is an information gradient producing a force by which the low energy system controls the high energy system, through a flow of information against the traditional energy gradient. Quantifying the information flow from a more complex system, high in information, to a less complex system, low in information, was the first tentative articulation of a universal principle of network science; the Principle of Complexity Management (PCM) [352]. The PCM was proposed as a quantitative statement of Wiener's more qualitative proposition and was proven for ergodic networks [10, 11], as well as, for non-ergodic networks [248]. Both proofs used non-equilibrium statistical physics to determine that the PCM entails information flow, from networks of high complexity, to those of low complexity, consequently exerting a force of one network on the other [363].

One definition of complexity, the one adopted in this essay, is that of a balance between regularity and randomness [347]. Consequently, complexity can be enhanced by increasing the number of degrees of freedom and/or the level of nonlinearity describing a system's dynamics. Both the increase in nonlinearity and the size of the system, increase the level of uncertainty in the Newtonian dynamics, resulting in an accompanying increase in the system's complexity, as measured by information. Large systems, made up of an interacting network of subsystems, are invariably heterogeneous, giving rise to information transport, in an attempt to reduce the disparity in complexity, minimize the information imbalance and maximize the entropy of the overall network.

Historically the disciplines of statistical physics and thermodynamics were thought to be sufficient for describing complex physical phenomena solely through the use of analytic functions. The recognition of the importance of scaling and fractals in the last half of the twentieth century, established that to understand complexity, science must go beyond the analysis of analytic functions, not just in physics, but in the social, ecological and life sciences, as well. This is the approach we take to understanding the origin of allometry relations within complex systems.

## 1.4 Measures of size

In this essay, I argue that size and complexity go hand-in-hand and are inextricably intertwined. They increase and decrease together, but not in direct proportion to one another. For example, as an organization grows it provides increasing opportunity for variability, independently of the function of the organization, and this variability is one measure of complexity, but not the only measure. Similarly, as the complexity of an organization of a given size increases, the organization's behavior-modes saturate and cease to vary, forcing the organization to either further increase in size, or failing to increase, to develop resistance to increasing in complexity. On the other hand, as an organization increases in size, it must also increase in complexity, which is necessary in order to maintain stability. The idea of a large, but simple, organization is an illusion. A small perturbation would cause such an organization to become unstable and to fail in any attempt to reorganize and adapt to a changing environment.

Another way to view the relation between complexity and size is by relating the functionality of the phenomenon of interest to its size. We shall show that the more sophisticated the functionality, the greater the complexity necessary to carry out that function. Consequently, we interpret the many empirical relations, between functionality and size, as being the result of the intrinsic relation between complexity and size, with the complexity being manifest through functionality. This subtle, yet ubiquitous, interrelation among complexity, functionality and size, is here hypothesized to form the foundation of the science of allometry.

We have used social organization as one exemplar of what has been determined to be a ubiquitous relation between size and complexity. In addition to the social realm, there is the physical domain in which an object's size is limited by its internal composition, the ecological realm in which the size of the ecosystem is determined by the complexity of the interactions among species and is a balance between dynamics and randomness, as well as, other disciplines in which aggregates are determined by the interdependence of size and complexity. In all these disciplines, size is determined by what can be quantified, even though the definition of complexity remains elusive. It is this elusive nature of complexity that suggests why functionality is the empirical measure that is related to size.

What is quantified is determined by that which we can experience; what we can see, smell, hear, feel and taste. But do not forget one of the intangibles, that being time, which we also directly experience, but in a different way. An elephant is one of the largest land animals on the planet and because of its massiveness we expect it to lumber, which it does. It lives longer than humming birds and other small animals, its heart beats more slowly, the rising and falling of its lungs is more measured, and it roams over larger territories. From birth, the lumbering elephant keeps time using a different clock than that used by the smaller animals. The size of the individual determines their experience of time, with which their life unfolds; a condition that would have made Sir Isaac uncomfortable.

Size is measured in a number of different ways, including, but not limited to, geometric dimensions. A plot of land is expressed in acres; income is measured in the currency of the country of interest; age can be quantified by the number of years since birth; mass is expressed in terms of a number on a weight scale, and so on. The size of the phenomenon of interest is always given by a number of fundamental units. In this way if $A$ is greater than $B$, and $B$ is greater than $C$, then it must be true that $A$ is greater than $C$. This simple ordering property of numbers, allows us to order the phenomena measured, in the same way.

However, this way of looking at things excludes a large slice of human experience. For example, suppose $A$, $B$ and $C$ are political candidates. If Jack and Jill are on the ballot; one voter's preference might be Jack. If Jill and John are on the ballot; that same voter's preference might Jill. This ordering does not, however, imply that if Jack and John were on the ballot that this voter would prefer Jack over John. The logic of simple quantification does not always map onto the analysis of people's decision making in a one-to-one way. Decision making does not always follow a simple ordinal logic. In fact, decision making consists of at least two distinct parts; the rational and the irrational. The rational is a relatively slow process of conscious thought. The irrational is the impulsive and much more rapid process of jumping to conclusions or making an immediate inference. For most people daily life is dominated by the fast rather than the slow process of decision making.

Consequently, not all phenomena, in which we are interested, can be quantified, even though we talk to one another as if they could. For example: what does it mean to say that Jack is happier than Jill? Happiness is not a simple process to which we can assign a single number; it contains positive feelings, multiple thoughts, a variety of physical responses and a myriad of other things. One person, having received a promotion is happy, with a warm glow, experiencing a positive attraction towards other people and may choose to share this state with others. Another individual may have just solved a difficult problem and is awash in the satisfaction that comes with engaging and overcoming a challenge. She may, or may not, want to share this experience with a loved one, or with a perfect stranger. Archimedes shared his discovery of the principle of buoyancy by leaping from his bath and running through town, naked, shouting Eureka, or so the story goes. The quiet satisfaction of the introvert, on the one hand, cannot be directly compared with the wild exuberance of the extrovert, on the other. Or if they can be compared would it be along the ordinal scale of quantification?

With this cautionary note in mind let us return to the quantification of size, keeping in the back of our minds that we also want to order processes like happiness, which, at least superficially, cannot be quantified.

### 1.4.1 Physical

Galileo Galilei (1564–1642) was the first of the modern scientists to appreciate the significance of size, and to record his insights in a publication. He recognized that in order for an organism, or physical structure, to retain a constant function, as its size increases, requires its shape (architecture), or the materials with which it is constructed, to change [109]:

> From what has already been demonstrated, you can plainly see the impossibility of increasing the size of structures to vast dimensions either in art or in nature...; so also it would be impossible to build up the body structures of men, horses, and other animals so as to hold together and perform their normal function if these animals were to be increased enormously in height; for this increase in height can be accomplished only by employing a material which is harder and stronger than usual, or by enlarging the size of the bones, thus changing their shape until the form and appearance of the animals suggest a monstrosity.

The connection between growth and form was also observed by D'Arcy Thompson [317] early in the previous century, when he explored problems of scale, size, and shape in the biological sciences. A simple example of Thompson's approach is provided by the determination of the maximum size of terrestrial bodies (vertebrates). The strength of a bone, in the simplest model, increases in direct proportion to its cross-sectional area (the square of its linear dimension), whereas the bone's weight increases in proportion to its volume (the cube of its linear dimension). Thus, there comes a point where a bone does not have sufficient strength to support its own weight, as first observed by Galileo in 1638. The point of collapse is given by the intersection of a quadratic curve denoting the strength of a bone and a cubic curve denoting its weight, cf. Figure 1.5.

The same reasoning applies to the design of bridges, as well as other physical structures. The architecture is such that the bridge must not only be able to support its own weight, but the additional weight of any traffic using the bridge, as well. Bridge

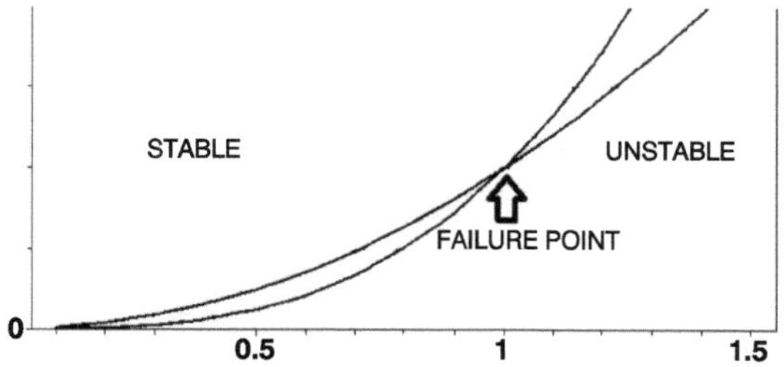


**Figure 1.5:** The strength of a bone increases with the cross-sectional area $A \sim l^2$, whereas its weight increases as the volume $W \sim l^3$. The two curves intersect at $A = W$. Beyoind this point the structure becomes unstable, the bone breaks and collapses under its own weight.

design has evolved over the millennia, from the known stone structures of the Roman Empire, to the modern suspension bridge of steel cables and girders. As the design changed to carry heavier loads, so too did the materials necessary to accommodate the design: stone to steel.

A second example, which is actually a variant of the first, also recognizes that mass increases as the cube of its linear dimension, but the surface area increases only as the square. Consequently, if one animal is twice the height of another, it is likely to be eight times heavier. We can immediately infer how the larger plants and animals compensate for their bulk through respiration. Note that respiration depends on surface area for the exchange of gases, as does cooling by evaporation from the skin and nutrition by absorption through membranes. One way to add surface to a given volume is to make the exterior more irregular, as with branches and leaves on trees; another is to hollow out the interior as with some cheeses. The human lung, with 300 million air sacs, approaches the more favorable ratio of surface to volume enjoyed by our evolutionary ancestors, the single-celled microbes [339].

In the physical domain the behavior of the individual elements do not change in a given phase (state) of matter. However, as a material transitions from one phase to another, say from liquid to solid, the properties of the constitutive individuals do change. The number of elements in a physical system is typically so large (infinite) that thermodynamics forces replace Newton's equations of motion for the point particles and the material properties are determined by the dynamics of the thermodynamic (average) variables. This effective infinite-size limit does not exist in social phenomena, there are always finite size effects.

But it is not just the size of the individual, but the size of the group to which an individual belongs that determines the individual's characteristics. People who live in large cities move faster and are more creative than those that live in small towns; earthquakes along with brainquakes, appear in bursts over time; stock market crashes are much more frequent than anticipated by professionals throughout the last fifty years; these and many more disjoint complex phenomena, all of which as we shall discuss, share a common origin. The source is not a mechanistic one, since the mechanisms generating the processes are distinct, rather the origin of the relation between the property of interest and the size of the phenomenon, manifesting that property, is complexity itself. In this sense the surrogate measure of complexity is functionality, through their mutual dependence on size.

### 1.4.2 Fractals

Benoit Mandelbrot [201, 202] identified multiple allometry relations (ARs), masquerading under a variety of empirical 'laws', and argued that they were a consequence of complex phenomena not having characteristic scales. He coined the term fractal and championed its use in all manner of social and natural phenomena. Subsequent interpretations of ARs often involve fractals and so we recall some fundamental properties of fractals that turn out to be useful for the present discussion. An intuitive definition of a geometrical fractal, given by its inventor [203] is:

> A fractal is a shape made of parts similar to the whole in some way.

The fractal concept is a good deal more subtle than this simple statement reveals at first glance. In fact, it arises in three distinct, but related guises; geometrical, statistical and dynamical. Geometric fractals deal with the self-similarity of complex geometric forms. A fractal object examined with ever increasing magnification reveals ever greater levels of detail; detail that is self-similar in character. The idea of self-similarity entails the repetition of a motif across all scales. The basic mathematical properties of geometric fractals and their myriad of applications can be found in a number of excellent books, e.g., [25, 97, 99, 201, 203, 214, 346], and are not repeated here. We merely record and interpret those properties that are later found to be useful.

Following the wisdom of the expression, "a picture is worth a thousand words", let us illustrate a fractal curve by seeing what goes into its construction. One kind of fractal curve is built starting from a straight line of unit length, dividing it into three equal sized sections and removing the middle third. The empty segment is replaced with an equilateral triangle having sides of length of the size of the segment being replaced, as is depicted in the top panel of Figure 1.6. This process is repeated on each of the line segments of the curve as shown in the second panel. This process is repeated indefinitely each time, with the same contraction of scale size, and proportional increase in the length of the curve.

It is certain, by mathematical construction and visual intuition, that what geometric fractals, such as the Koch curve in Figure 1.6, have in common, is a repeating pattern at every scale. If the replication is exactly the same at every scale, then the curve is self-similar.

**Figure 1.6:** The generation of Koch curve. In the limit of infinite application of the generation process the length of the curve becomes infinite.

As Latka et al. [181] point out, the self-similar pattern of a fractal curve can be viewed from another perspective by examining a different geometrical fractal. In the left column of Figure 1.7 we observe a branching tree in which the outermost 'tip' of the branch is magnified to reveal the same branching structure at the next smaller scale. Such branching may be found in the architecture of the human lung and vascular system [357]. A mathematical fractal has no characteristic scale size and its defining pattern proceeds to ever smaller and ever larger scales. On the other hand, a natural fractal always terminates at some smallest and largest scale and whether or not this is a useful concept, for the process considered, depends on the extent of the interval over which the process appears to be scale-free. A rule of thumb is that if, the scale-free character persists over two orders of magnitude, then the fractal concept may be useful.

However, it is not just in modeling patterns in space that fractals have revealed their usefulness, but also for the time intervals between events. The right hand column in Figure 1.7 depicts the time series for the time interval between beats of the human heart for a healthy youth. The uppermost panel shows such a time series for 300 minutes. The blocked region of 30 minutes is then magnified to provide the second panel moving downward from the top. It is determined that the time series in the first and second panels have the same statistical distribution, which is to say that the two time series are statistically self-similar. Going to smaller and smaller scales, the statistical self-similarity parallels the geometrical fractal behavior seen on the left side of the figure.

One way to characterize a fractal object is by means of its dimension. In the world of Euclid a point has zero dimension, a line is one-dimensional, a surface has two dimensions, and so on. So what does it mean for an object to have a non-integer di-

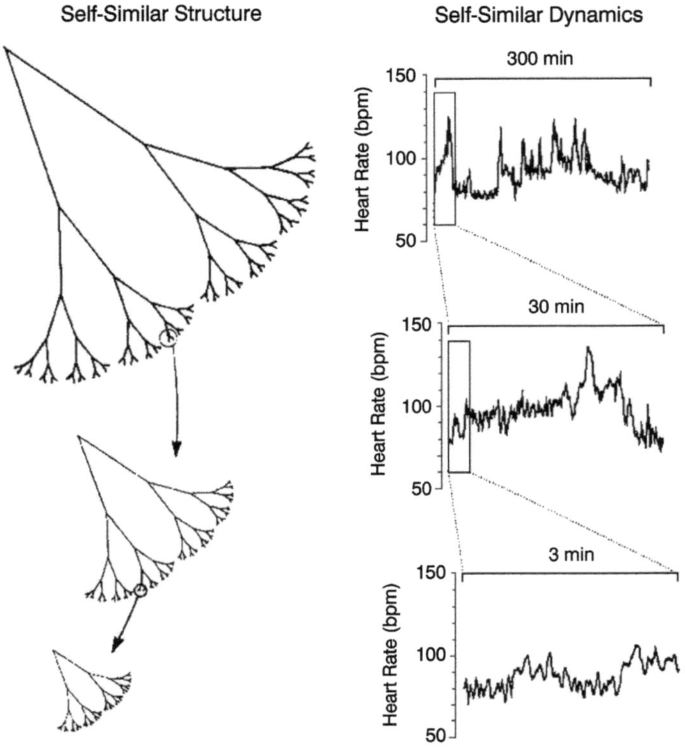


**Figure 1.7:** Geometric fractals are a family of shapes containing infinite levels of detail, such as shown by the fractal tree on the left. Statistical fractals have the same kind of self-similarity, but in terms of the statistical fluctuations. The repeating shapes arise in the time series depicted on the right.

mension? We determine dimension by means of measurement. The length of a curve $L$ is determined by laying down a ruler of size $r$, a given number of times $N$, such that

$$L = Nr. \tag{1.1}$$

Similarly, the area $A$ of a surface is determined by how many times $N$, a unit area $r^2$, is required to cover it

$$A = Nr^2. \tag{1.2}$$

Thus, we may say that the number of self-similar objects $N$ required to cover an object of dimension $D$ is given by

$$N \propto r^{-D}, \tag{1.3}$$

where $r$ is the size of the 'ruler'. In this way the fractal dimension of the object being covered can be mathematically defined as

$$D = -\ln N / \ln r, \tag{1.4}$$

in the limit of vanishing $r$. As the ruler size goes to zero, the number of elements (unit line segment, unit area, unit volume, etc.) necessary to cover the object diverges to infinity, in such a way that $D$ remains finite for self-similar objects. This dimension is not necessarily integer valued, which is the point of the discussion.

The fact that the traditional mathematical definition of dimension does not require integer values is interesting. Of course, historically the non-integer values were systematically dismissed by most physical scientists, arguing that they are not physically observable. That is where the matter stood until Mandelbrot showed that such non-integer values consistently appear in a number of complex physical phenomena, for example, in turbulent fluid flow [201]. Once the physical scientists began to concede the possible utility of fractals, it did not take long for investigators in the social and life sciences to recognize that fractal processes had been dominating their complex worlds all along. For over a quarter century scientists have cataloged phenomena that have fractal dimensions in time series, statistics and in all manner of variables. These dimensions were systematically shown to be related to the concept of scaling.

An observable $Z(t)$ is scaling if, for a positive constant $c$, it satisfies the homogeneity relation

$$Z(ct) = c^H Z(t). \tag{1.5}$$

Modifying the units of the independent variable $t$ therefore only changes the overall observable by a multiplicative factor; this is self-affinity. Barenblatt [20] remarked that such scaling laws are not merely special cases of more general relations; they never appear by accident and they always reveal self-similarity. Note that scaling alone is not sufficient to prove that a function is fractal, but if a function is fractal it does scale.

We subsequently relax the distinction between self-affine and self-similar, since self-similarity has been informally extended to encompass both meanings in the physics literature. Scaling requires that a function $\Phi(X_1, \dots, X_N)$ be such that scaling each of the $N$ variables by an appropriate choice of exponents $(\alpha_1, \dots, \alpha_N)$ always recovers the same function up to an overall constant:

$$\Phi(X_1, \dots, X_N) = \gamma^\beta \Phi(\gamma^{\alpha_1} X_1, \dots, \gamma^{\alpha_N} X_N). \tag{1.6}$$

We observe that equation (1.5) is the simplest of such scaling relations, between two variables, such that they satisfy the renormalization group relation

$$Y(\gamma X) = \gamma^\beta Y(X). \tag{1.7}$$

The lowest-order solution to this equation is obtained using the same procedure used to solve differential equations: guess the answer, plug it into the equation of interest and determine the unknown quantities necessary to satisfy the equality.

To solve equation (1.7), we assume a solution of the form

$$Y(X) = \frac{A(X)}{X^{\mu}}, \tag{1.8}$$

which when substituted into equation (1.7) generates the conditions on the assumed form of the solution

$$\mu = -\beta; \quad \text{and} \quad A(\gamma X) = A(X). \tag{1.9}$$

The conditions specified by equation (1.9) require $A(X)$ to be a periodic function of $\ln X$, with period $\ln \gamma$. Consequently, the general solution, which we subsequently discuss in some detail, has the form of an infinite series of log-periodic terms

$$A(X) = \sum_{k=-\infty}^{\infty} A_k X^{\frac{i2\pi}{\ln \gamma}k} = \sum_{k=-\infty}^{\infty} A_k \exp\left[ i2\pi k \frac{\ln X}{\ln \gamma} \right] \tag{1.10}$$

The lowest-order solution has $\alpha = A(X) =$ constant, such that equation (1.8) becomes

$$Y = \alpha X^{\beta}, \tag{1.11}$$

which has the form of an allometry relation (AR), where changes in the network size $X$, control (regulate) changes in the functionality $Y$ in living networks. This type of regulation is also common in physical and social networks through precisely this kind of homogeneous scaling relation. We shall have a great deal more to say regarding scaling, fractals and ARs.

Inhomogeneity in space and intermittency in time are the hallmarks of fractal statistics and it is the statistical, rather than the geometrical, sameness that is evident at increasing levels of magnification in the class of complex phenomena we investigate. In geometrical fractals the observable scales from one level to the next. In statistical fractals, where the phase space variable $z$ and trajectory, parameterized with the time $t$, replace the dynamic variable $Z(t)$, it is the PDF, $P(z, t)$ that satisfies a scaling relation

$$P(\alpha z, \beta t) = \beta^{-\mu} P(z, t); \quad \mu = \ln \alpha / \ln \beta, \tag{1.12}$$

and the homogeneity relation of equation (1.7) is interpreted in the sense of the PDF in equation (1.12). Time series, with such statistical properties, are found in multiple disciplines including finance [204], allometry [351, 360], economics [205], neuroscience [6, 335], geophysics [323], physiology [348] and general complex networks [352]. An extensive discussion of statistical data, with such scaling behavior, is given by Beran [34] in terms of the long-term memory captured by the scaling exponent. One example of a scaling PDF is given by

$$P(z, t) = \frac{1}{t^{\mu}} F_z\left( \frac{z}{t^{\mu}} \right), \tag{1.13}$$

and in a standard diffusion process $Z(t)$ is the displacement of the diffusing particle from its initial position at time $t$, $\mu = \frac{1}{2}$ and the functional form of $F_z(\cdot)$ is the Gauss PDF. However, for general complex phenomena there is a broad class of distributions for which the functional form of $F_z(\cdot)$ is not bell-shaped and the scaling index $\mu \neq \frac{1}{2}$, see Chapter 3 for additional discussion. It is subsequently shown that the scaling PDF is the general solution to a fractional phase space equation of evolution.

Dynamic fractals do not directly enter our discussion, however, for completeness, we mention that in a dynamic fractal the geometry of the manifold on which the dynamics of a network unfolds is fractal, so that the associated chaotic time series is also fractal [236].

### 1.4.3 Fractal advantage

Why are fractals important in the design of allometry systems (networks)? Barenblatt and Monin [21] suggested that metabolic scaling might be a consequence of the fractal nature of biology and it has been determined by a number of investigators that fractal geometry maximizes the efficiency of nutrient transport in biological networks [365]. Weibel [331] maintains that the fractal design principle can be observed in all manner of physiologic networks quantifying the observations and speculations of Mandelbrot [201, 202], as well as, those of West [346].

West [340, 341] presented a simple argument establishing that fractals are more adaptive to internal changes and to changes in the environment than classical models of processes and structures would suggest. Consider a network property characterized by classical scaling at the level $k$, such as the length or diameter of a branch is given by

$$F_k \propto e^{-\lambda k}, \tag{1.14}$$

compared with a fractal scaling of the same property

$$F_k \propto k^{-\lambda}. \tag{1.15}$$

What is significant about these two functional forms for the present argument is the dependence on the parameter $\lambda$. The exponential has emerged from a large number of optimization arguments, using the differential calculus, and the IPL results from the renormalization group theory (RGT) scaling arguments.

Assume the parameter $\lambda$ determines the stability of the process being examined. The parameter is taken to be the sum of a constant part $\lambda_0$ and a random part $\xi$. The random part can arise from unpredictable erratic changes in the environment during morphogenesis, non-systematic errors in the code generating the physiologic structure, or any of a number of other causes of irregularity. Thus, regardless of whether the errors are induced internally or externally, the average is taken over an ensemble of zero-centered Gaussian fluctuations $\xi$ with variance $\sigma^2/2$. Note that the choice of Gauss statistics has no special significance here except to provide closed form expressions for the averages to facilitate discussion. The relative error generated by the fluctuations is given by the ratio of the average value, denoted by the brackets, to the function in the absence of fluctuations, yielding the relative error for classical scaling

$$\varepsilon_k^{(\text{normal})} = \frac{\langle e^{-k(\lambda_0+\xi)}\rangle}{e^{-k\lambda_0}} = \exp[\sigma^2 k^2] \tag{1.16}$$

and for fractal scaling

$$\varepsilon_k^{(\text{fractal})} = \frac{\langle e^{-\ln k(\lambda_0+\xi)}\rangle}{e^{-\ln k\lambda_0}} = \exp[\sigma^2(\ln k)^2]. \tag{1.17}$$

In both expressions $\sigma^2$ provides a measure of the strength of the fluctuation-induced error.

The two error functions are graphed in Figure 1.8 for fluctuations with a variance $\sigma^2 = 0.01$. At $k = 15$ the error generated by classical scaling is 9.5. This enormous relative error implies that the perturbed average property at generation 15 differs, by nearly an order of magnitude, from what it would be in an unperturbed network. A biological network with this sensitivity to error would not survive in the wild. For example, the diameter of a bronchial airway in the human lung could not survive this level of sensitivity to chemical fluctuations during morphogenesis. However, that same property in a fractal network only changes by 10% at the distal point $k = 20$. The implication is that the fractal network is relatively unresponsive to fluctuations in the scaling process.

A fractal network is consequently very tolerant of variability. This error tolerance can be traced back to the broadband nature of the distribution in scale sizes of a fractal object. This distribution ascribes many scales to each generation within the net-

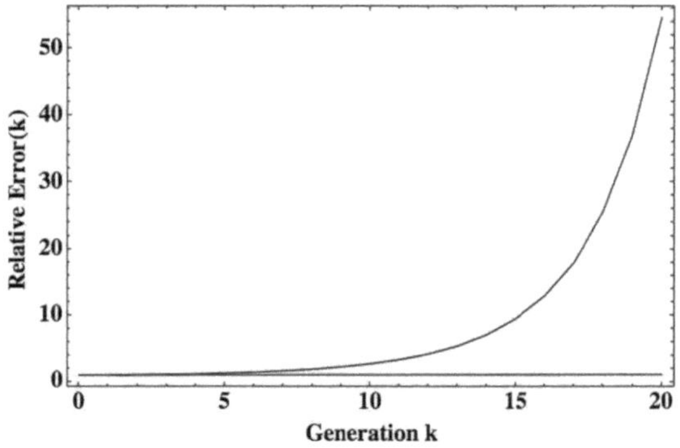


**Figure 1.8:** The error between the model prediction and the prediction averaged over a noisy parameter is shown for the classical model (upper curve) and the fractal model (lower curve).

work. The scales introduced by the errors are therefore already present in a fractal object. Thus, the fractal network is *preadapted* to variation and is therefore insensitive to change [340, 341], thereby affording a corresponding survival advantage. These conclusions do not vary with modification in the assumed statistics of the errors and consequently we would expect to see this kind of adaptability in all manner of living and social networks. Thus, we anticipate that physiologic networks, such as the bronchial airways in the lung, and the body's nutrient transport network, among many others, are fractal.

It cannot be emphasized too strongly that it is the statistical fractal that confers a stability advantage to an organism adopting this strategy. In this sense complex adaptive behavior does not require a mechanism-specific response to environmental fluctuations in order to maintain fitness; scaling statistics achieves this purpose.

## 1.5 Allometry heuristics

While the statistics of data analysis was being introduced by mathematicians to interpret experimental uncertainty, another remarkable contribution to science was being made by the zoologist Curvier [76]. He empirically determined that the brain mass of a mammal increases more slowly than does its total body mass (TBM). Cuvier's observations constituted the initial heuristic study of allometry, which relates the size of an organ to the size of the host organism, with mass serving as a measure of size. If the size of the brain mass is denoted by $Y$ and the TBM by $X$, then Curvier's experiments can be summarized by the schematic AR:

$$Y = aX^b, \tag{1.18}$$

in terms of the allometry coefficient $a$ and the allometry exponent $b$. A more general AR is one in which $Y$ is not the size of an organ, but instead is a functionality of the organism, but the size of the organism is still characterized by $X$. Subsequently, ARs have been uncovered in every scientific discipline from Astronomy to Zoology, with a system's functionality given by $Y$, the system's size by $X$ and the allometry parameters $(a, b)$ characterizing the system. Note that the AR is given by a solution to the renormalization group relation, equation (1.7).

An empirical AR is obtained by means of simultaneous measurements of a system's functionality and size to form the data pair $(Y, X)_t$ and $t$ is the time of the measurement. Note that the schematic AR given by equation (1.18) is typically interpreted as being independent of time, so that in practice the variables are replaced by their static average values. Denoting an average over the data pair by a bracket, assuming a very large number of measurements for a system in a steady state, yields

$$\langle Y \rangle = a \langle X \rangle^b, \tag{1.19}$$

which is the heuristic AR found in applications. Focusing on the statistical nature of allometry, we emphasize that ARs ought to be strictly a relation between average quantities [57, 288, 355], not a relation between the dynamic variables.

It must be borne in mind that the AR does not imply a causal relationship. The mathematicians, Taskinan and Warton [312] make the following observation:

> The lines fitted to data in allometric studies are not for prediction of one variable from the other, rather they are lines fitted to summarize the relationship between the two variables, and to do so in a manner that is symmetric with respect to the $X$ and $Y$ variables.

Gayon [112] reviewed the history of the concept of allometry, defined as the study of body size and its consequences [128, 267] within a given organism and between species in a given taxon. He distinguished between four different forms of AR: (1) ontogenetic allometry, which refers to relative growth in individuals; (2) phylogenetic allometry, which refers to constant differential growth ratios in lineages; (3) intraspecies allometry, which refers to adult individuals within a species; (4) interspecies allometry, which refers to the same kind of phenomenon among related species. However, the theoretical entailment of static from dynamic allometry models has not been systematically studied, although there has been some recent effort in that direction [349, 351] and that we subsequently discuss.

### 1.5.1 Allometry/information hypothesis

Given the variety of disciplines in which ARs have been found [355], any fundamental principle on which to base allometry behavior must be discipline independent, or more specifically it must be independent of any mechanism that is characteristic of a given discipline. We have argued elsewhere [353, 355] that the empirical AR given by equation (1.19) is a consequence of the imbalance between the complexity associated with the system functionality and the complexity associated with the system size, both being measured by Shannon information. We refer to this as the allometry/information hypothesis (A/IH) [363] and postulate that in a complex network, composed of two or more interacting sub-networks, the flow of information is driven by the complexity gradient between the sub-networks, from that with the greater to that with the lesser complexity. Thus, according to the Principle of Complexity Management (PCM) this imbalance produces an information force [363] that entails an AR within the complex network.

Implicit in the A/IH is the assumed existence of dependencies between complexity and both system size and system functionality. Such dependencies have been observed in: the positive feedback between social complexity and the size of human social groups [72], as well as, in ant colony size [104]; and the increase in biological complexity with ecosystem size [56]. Other relations have been observed in multiple disciplines, including the increase of prey refuge from predators with habitat complexity [127]; computational complexity increasing with program size [159]; and gene functionality depending on system complexity [154]. We abstract from these observations that the complexity of a phenomenon increases with system size and that the system functionality increases with system complexity.

In order to prove our hypothesis requires introducing a number of new concepts, not the least of which is a new kind of calculus. This 'new' form of calculus is not really new, since it has existed since Leibniz, but it is relatively new to the mainstream scientific community. This is the fractional calculus, which involves the manipulation of fractional-order differential and integral operators. Our purpose is not to discuss, with any mathematical rigor, the fine points of the fractional calculus, but rather to use some of its more well-known properties to prove that ARs are a consequence of spatial inhomogeneity and temporal anisotropy of the statistical behavior of complex systems.

We take up the fractional calculus after we have discussed a bit more about how information is passed back and forth between complex sub-networks in a complex host network.

### 1.5.2 Size variability

An AR equation interrelates two observables in a complex phenomena, say $X_{ij}$ and $Y_{ij}$. Here we anticipate a biological or ecological application, where the first index $j$ denotes an individual and the second index $i$ denotes a species. In living networks this measure of size is typically TBM and the schematic intraspecies AR is

$$Y_{ij} = aX_{ij}^{b}. \tag{1.20}$$

Again, by convention, the variable on the right in the theoretical AR is taken to be the measure of size and that on the left to be the measure of functionality. To consider the variability in allometry data let us stay focused, for the moment, on biological systems, but the same general considerations can be readily adjusted to apply to social and ecological data as well.

Gould [128] stressed that allometry laws in the life sciences fall into two distinct groups. The *intraspecies* AR relates a property of an organism within a species to its TBM $M_{ij} = X_{ij}$ in equation (1.20). The *interspecies* AR relates a property across species such as the average basal metabolic rate (BMR $B_{ij}$) to average TBM [57, 288]. The average size of an adult member of the species $i$, such as average TBM of a species of lobster, deer, bird, etc., is

$$\langle M_i \rangle \equiv \frac{1}{N}\sum_{j=1}^{N} M_{ij} \tag{1.21}$$

and the average functionality for the species $i$, such as average BMR, is

$$\langle B_i \rangle \equiv \frac{1}{N}\sum_{j=1}^{N} B_{ij} \tag{1.22}$$

so that the empirical interspecies AR is written in general as

$$\langle B_i \rangle = a\langle M_i \rangle^b. \tag{1.23}$$

The two kinds of AR, given by equations (1.20) and (1.23), are distinctly different from one another and the models developed to determine the theoretical forms of the allometry coefficient $a$ and allometry exponent $b$, in the two cases, are quite varied. Note that both ARs are traditionally expressed with the indices suppressed, so that both $M_{ij}$ and $\langle M_i \rangle$ are usually written as $M$, resulting in confusion between the two forms of AR.

Savage [282] points out that what we are faced with here is the well-known *fallacy of averages*, which is the fallacy of replacing a variable by its average value in any expression in which the variable is not linear. In the AR this means that in general

$$\langle M_i^b \rangle \neq \langle M_i \rangle^b, \tag{1.24}$$

except in the singular case $b = 1$. He goes on to calculate the differences in the two averages for a variety of metabolic rate data, using $b = 3/4$, and determines that errors between 8% and 54% occur, depending on the data set used. He concludes that the substitution of a variable by its average is acceptable under some well-defined conditions, but is in general not valid in allometry studies, as one might expect.

Equations (1.20) and (1.23) look very much like the scaling relations that have become so popular in the study of complex networks over the last decade, see, for example, [4, 53, 231, 328, 347]. Historically the nonlinear nature of these equations, $b \neq 1$, has precluded their direct fitting to data. However, the analysis of real data is traditionally done by taking the logarithm of the data and in doing a linear regression on intraspecies data of the equation

$$\ln B_{ij} = \ln a + b \ln M_{ij} \tag{1.25}$$

or on interspecies data of the equation

$$\ln\langle B_i \rangle = \ln a + b \ln\langle M_i \rangle. \tag{1.26}$$

Note that these are two independent ways of treating data and they cannot be related to one another in general.

Linear regression analysis focuses on the conditional PDF of $B$, given $M$, and is often used to quantify the strength of the relation between the two variables, or for forecasting. This is the interpretation that is often implicitly assumed in the data analysis to determine the AR. However the fact that $M$ and $B$ are measured independently indicates that this interpretation of linear regression is not appropriate for the data analysis using equation (1.25) or (1.26). The independent measurements suggest that it is more appropriate to address the joint PDF for bivariate analysis of the data. We explore this and related statistical questions in Chapter 3.

### 1.5.3 On AR models

D'Arcy Wentworth Thompson began and ended his seminal book *On Growth and Form* [317] arguing the need for more mathematics in the understanding of the *Natural Sciences*. As pointed out to me by a colleague, this is the centenary of the first edition of his book, and there is a certain satisfaction in documenting how far the field has come in realizing his goal. Thompson opened his work with a penetrating discussion of the *Principle of Similitude*, reviewing dimensional analysis and dimensionless constants in biology, thereby laying the groundwork for his discussion of the growth of organisms. In his second chapter, he connects his scaling arguments to those of Huxley; the latter following the compound interest law. In the 1942 edition of his book, with his unfailing prescience, Thompson expressed skepticism as to the generality of the law proposed by Huxley.

Sir Julian Huxley [152], a member of a family famous for their multiple contributions to theoretical biology, proposed that an organ, within an organism, has a rate of growth proportional to the size of the organism. This was the first dynamic 'theory' proposed to explain the form of the intraspecies AR. Huxley suggested that if $Y_{ij}$ is from a living subnetwork observable from the $j^{\text{th}}$ member of the $i^{\text{th}}$ species, with growth rate $\gamma_i$ and $X_{ij}$ is a measure of the size of the host organism, from the $j^{\text{th}}$ member of the $i^{\text{th}}$ species, with growth rate $\vartheta_i$, then the fractional increase in the two are equal:

$$\frac{1}{\gamma_i}\frac{dY_{ij}}{Y_{ij}} = \frac{1}{\vartheta_i}\frac{dX_{ij}}{X_{ij}}. \tag{1.27}$$

This equation can be directly integrated to obtain the time-independent intraspecies AR given by equation (1.20), where $a$, a constant of integration, and $b(=\gamma_i/\vartheta_i)$, determined empirically from the observed growth rates.

Huxley's argument may appear somewhat naïve today, but it must be borne in mind that the intraspecies AR is an odd form of relation. By odd we mean that the quantities, on the two sides of equation (1.20), are independently measured and the form of the AR is assumed, based on empirical evidence. Even the form of the logarithmically transformed equations, which are used to fit the allometry parameters to data, using least-square error analysis, are often misinterpreted in terms of dependent and independent variables. Consequently assuming a dynamic equation of the form equation (1.27) to underlie allometry has dramatic biological implications for growth [152].

Modern explanations of AR begin with the application of either fractal geometry, or fractal statistics, to the interpretation of scaling phenomena. The detailed application of fractal geometry to the explanation of intraspecies ARs is a little over a decade old and, although well received, has not been universally accepted. An alternate perspective is given by the interspecies AR based on linear regression analysis of fluctuating data sets as discussed in Chapter 2. We again emphasize that the intraspecies and interspecies ARs are not the same and subsequently show that the interspecies AR can only be derived from the intraspecies one for a narrow distribution of fluctuations, a condition not often met by data. We also distinguish between two kinds of statistical models; those in which $X$ and $Y$ are statistical quantities, with $a$ and $b$ fixed; in contrast to those in which $a$ and $b$ are statistical quantities, with $X$ and $Y$ fixed. It should be emphasized that the functional form of the AR, including which quantities are deterministic and which are random, are questions to be dynamically answered. The empirical PDF of metabolic allometry coefficients is shown to be of the Pareto IPL form in Chapter 3.

A number of reductionist arguments converge on the conclusion that the allometry exponent is universal, and we review some of these arguments, subsequently. However, investigators have also derived a deterministic relation between the allometry exponent $b$ and the allometry coefficient $a$, some empirically and others using the fractional calculus. The covariation of the allometry parameters is clearly inconsistent with any assumption that the allometry exponent has a universal value. In Chapter 4 the discussions are organized to reach the conclusion that the interspecies physiologic AR is entailed by the scaling behavior of the PDF, which is derived using the fractional probability calculus.

We use the generic term network in our narrative in order to transition, with minimal confusion, from arcane historical theory to modern perspectives, using complex network theory. The mathematics of renormalization group theory (RGT) [162, 380], fractional differential equations (FDE) [195, 218, 251, 362], fractional stochastic differential equations (FSDE) [345] and transitioning from dynamic variables to phase space variables to express the probability calculus in terms of fractional diffusion equations [306, 345] are herein found to provide insight into different aspects of the origins of ARs.

## 1.6 Overview

We have reviewed some of the empirical links within complex phenomena, between size and function. A concrete example of such an empirical link, identified in biology nearly two hundred years ago, related the mass of an organ within an organism, such as the brain, to the organism's TBM. Grenfell et al. [132], among others, point out that biologists have described many such relationships linking body size to rates of physiological processes, interconnecting more than 21 orders of magnitude of TBM [211].

Over the course of time, such interdependency became known as allometry, literally meaning by a different measure and such links have been identified in nearly every scientific discipline. Allometry theory has acquired a mathematical description, through its many relations, along with a number of theoretical interpretations to account for its mathematical form. However, no one theory, has been accepted as successfully explaining the natural patterns captured by the ARs in their many guises, so the corresponding origins remain controversial. Consequently, in our subsequent review of the properties of allometry data, along with their various theoretical explanations, we provide a glimpse into those proposed origins.

All complex dynamical networks manifest fluctuations, either due to intrinsic nonlinear dynamics producing chaos [191, 236], or as the result of coupling of the network to an infinite dimensional, albeit unknown environment [184], or both. It should be emphasized that such statistical variability, along with the associated uncertainty, is the result of complexity and is independent of any question of measurement error. The modeling strategies adopted to explain ARs in the natural sciences have traditionally taken one of two roads: the statistical approach in which residual analysis is used to understand statistical patterns and identify the causes of variation in the AR [57, 267, 288]; or the reductionist approach to identify mechanisms that explain specific values of the allometry parameters in a particular discipline [18, 365]. We find that neither approach alone can provide a complete explanation of all the phenomena described by ARs. Therefore, we herein adopt a third approach using the fractional probability calculus to establish the validity of the A/IH, which we explain in due course.

In Chapter 2 we set the stage, with a brief history of empirical allometry. The observed exemplars are drawn from a wide range of disciplines, from inert physical networks, to the living networks of biology, ecology, physiological and sociology. The variability in the kinds of ARs presented, motivates the discussion of a ‘new’ kind of mathematics to model the different kinds of networks that give rise to the multiple ARs. The fractal nature of the underlying processes entails the use of fractional differential equations to describe their dynamics [195, 218, 251, 345] in Chapter 3. This approach to describing complexity also entails the transitioning from dynamic variables to phase space variables, to express the probability calculus in terms of fractional phase space equations (FPSEs) [166, 306, 345]. We emphasize here that there are a number of texts discussing the mathematical details of fractional differential equations [195, 216, 218, 251, 345, 362], but for our purposes it is sufficient to focus on the scaling behavior of the solutions to such equations. The scaling solution to the FPSE is found to provide insight into different aspects of the potential origins of allometry.

The probability calculus extended to fractional operators should enable modelers to associate characteristics of the measured PDF with specific deterministic mechanisms and with structural properties of the coupling between variables and fluctuations as we show in Chapter 4. The FPSE is presented as the basis for the many guises of allometry and their fractional form is based on a subordination argument through which the influence of the environment on the complexity is taken into account. Elsewhere we developed an alternative route to the probability calculus that systematically incorporates both reductionistic and statistical mechanism into the phenomenological explanation of ARs [349, 353]. Those arguments are extended in Chapter 5, using the fractional calculus. The exact asymptotic solutions to the fractional equations are used to directly establish the validity of the A/IH by calculating the empirical AR relating the average functionality to the average size.

# 2 Empirical allometry

This chapter briefly touches on various disciplines in which data have revealed patterns described by ARs and although not exhaustive, the examples presented demonstrate the wide range of application of allometry science. If allometry is to remain a fundamental area of scientific investigation and not be relegated to the status of an interesting curiosity from the Natural Sciences, it must have some degree of universality. One kind of universality would be as an AR with a specific value for a modeling parameter, such as the allometry exponent $b$ being 3/4 or 2/3, for a large class of phenomena [17, 366]. That is not the kind of universality envisioned here, for a number of reasons, one being that the specific value of the allometry exponent is not supported by all the data, as we show. This class of universality has been a long standing topic of controversy in physiology [129], in biology [89] and ecology [80] and many of the recent arguments are summarized in West and West [355].

The kind of universality we have in mind is a consequence of the existence of a generic framework for understanding the origin of ARs, independently of any specific mechanism, within any particular discipline. To motivate this second kind of universality we catalog a number of phenomenological ARs that are not usually discussed from a common perspective, a number of which are taken from the review [355]. In the present chapter we review and catalog a number of phenomenological relations, which are usually discussed from very different, discipline-specific, perspectives. There have been a large number of empirical relations that began with the identification of a pattern in data; were shown to have a terse power-law description; were interpreted using existing theory; reached the level of 'law' and given a name; only to subsequently fade away, when it proved impossible to connect the 'law' with a larger body of theory and/or data. Many such laws met their demise due to the inability to properly handle the statistical variability in the data. A failure we rectify in subsequent chapters.

We mentioned in Chapter 1 that Newton introduced the Principle of Similitude in the *Principia*, being fully cognizant of the fact that the law of universal gravitation was scale free. However, a century before his articulation of that principle, the empiricist Leonardo da Vinci recognized the existence of such scaling in his *Notebooks* [269]. He considered a branching tree depicted in Figure 2.1, which relates the diameter of a parent limb $d_0$ to two daughter limbs $d_1$ and $d_2$:

$$d_0^\alpha = d_1^\alpha + d_2^\alpha. \tag{2.1}$$

The da Vinci scaling relation supplies the phenomenological mechanism necessary for an AR to emerge in a number of disciplines, as we subsequently discuss.

There has been a resurgence of interest in the da Vinci empirical rule [94, 219], given by equation (2.1), which is interpreted to mean that the sum of the cross-sectional areas of tree branches, above a branching point, is equal to the cross-

DOI 10.1515/9783110535136-002

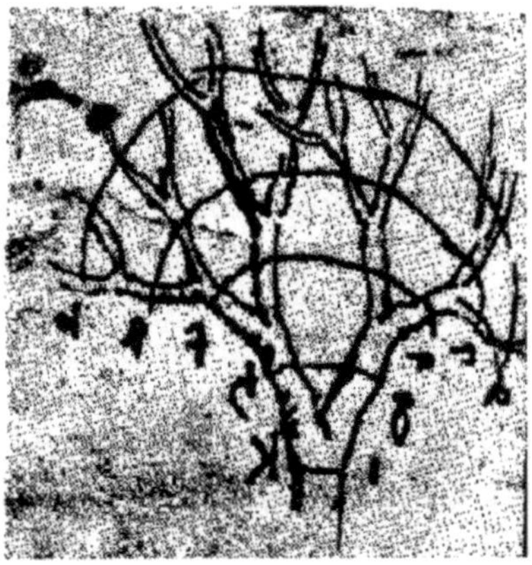

**Figure 2.1:** The sketch by da Vinci indicates daughter branches of a tree that have equivalent sizes, as measured by the diameter. Each semi-circle depicts a given generation of the branching process. (From da Vinci's *Notebooks* [269].)

sectional area of the branch below the branching point. Eloy [94] examines botanical trees, using theoretical arguments based on fractals, and establishes that if a tree skeleton has a self-similar structure and the branch diameters adjust to resist wind-induced loads, then da Vinci's rule results. Minamino and Tateno [219] also examine tree branchings and establish, using numerical simulations, that the da Vinci Rule is consistent, within certain limits, with a number of well-established biomechanical models.

Nearly five hundred years after da Vinci recorded his observations in Figure 2.1, Murray [228] used an energy minimization argument, which we review in Chapter 3, to derive an equivalent equation with the theoretical value $\alpha = 3$. The equation is known in the literature as the Murray Law or the Murray–Hess Law. In the simplest case the diameters of the daughter branches are equal $d_1 = d_2$, so that the da Vinci scaling relation reduces to a scaling between sequential generations of a bifurcating branching network having daughter branches of equal diameters:

$$d_{k+1} = 2^{-1/\alpha} d_k. \tag{2.2}$$

Such a scaling rule results in an exponential reduction in branch diameter, from generation to generation,

$$d_k = 2^{-k/\alpha} d_0 = d_0 e^{-k\lambda_\alpha}; \quad \lambda_\alpha = \frac{\ln 2}{\alpha} > 0. \tag{2.3}$$

Experiments have determined that the exponent $\alpha$ depends on the phenomenon being measured, as we shall see.

The above argument constitutes classical scaling theory, with a constant reduction of diameter between successive generations. In Section 1.4.3 we discussed the sensitivity of classical scaling to fluctuations in the growth process and showed how fractal scaling was relatively insensitive to such fluctuations. We here address some of the empirical evidence that nature provides in support of the suggested fractal advantage in the evolution of living networks. In the present chapter we do not distinguish between a dynamic variable and its average value in the notation used. We do this to avoid the endless use of qualifiers in the discussion, but we do emphasize the differences between random variables and their averages in subsequent chapters.

## 2.1 Living networks

Mature living networks can have static interspecies and intraspecies ARs, which are the structural vestige of a specific kind of growth process. The intraspecies ARs link two distinct, but interacting parts of the same organism in terms of mass, with TBM serving as a measure of size. Smith [299] maintained that concentrating on a power function as the method for evaluating the biological consequences of size has masked the complexity of the allometry problem. We agree with this observation, but perhaps in ways that Smith would not have anticipated, as will become evident. In addition the interspecies ARs link the functionality of an animal across species, but within a given taxon, with the average TBM of an adult member of the species serving as the measure of size.

### 2.1.1 Biology

The fact that brain mass increases more slowly than body size, as we proceed from small to large species within a taxon, was first recognized at the turn of the nineteenth century by Cuvier [76]. Almost a century passed before the empirical observation was first expressed mathematically as an AR by Snell [302]:

$$\text{brain weight} = a(\text{body weight})^b \tag{2.4}$$

where, on log–log graph paper, $a$ is the intercept with the vertical axis and $b$ is the slope of the line segment. Data supporting this AR has been steadily accumulating over the intervening centuries as shown in Figure 2.2.

Mammalian neocortical quantities $Y$ have subsequently been empirically determined to change as a function of neocortical gray matter volume $X$ as an AR given by equation (1.18). The neocortical allometry exponent was first measured by Tower [319] for neuron density to be approximately $-1/3$. Brains have another feature that leads to allometry scaling and that is surface area, with its multiple scale of foldings. The total surface area of the mammalian brain was found to have an allometry exponent of approximately $8/9$ [148, 156, 255]. Changizi [64] points out that the neocortex undergoes a complex transformation covering the five orders of magnitude from mouse to whale, but the ARs persist; those mentioned here along with many others.

A related AR is that between the volume of white matter and the volume of gray matter in the brain, as depicted in Figure 2.3. The scientist He [140] points out that the

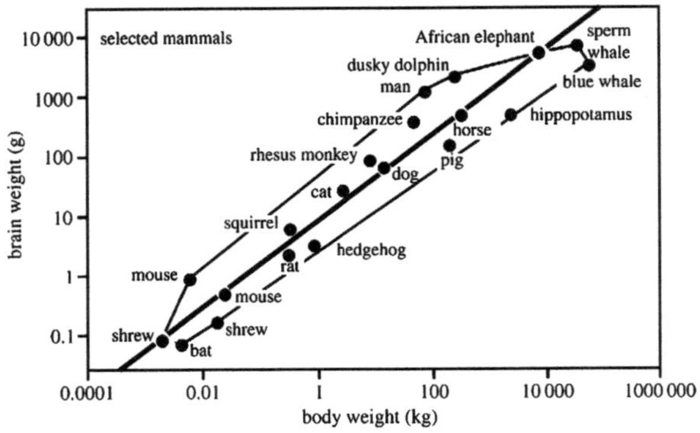


**Figure 2.2:** The brain weight versus body weight of animals spanning a range from shrew to sperm whale is graphed on log–log graph paper. The solid curve is the best fit of equation (2.4) to the data. (From [87] with permission.)

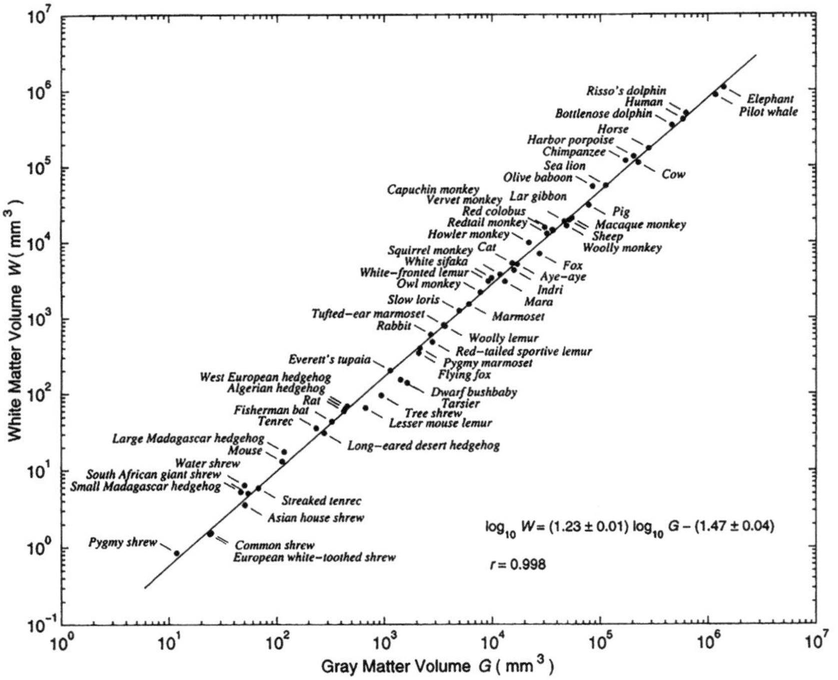


**Figure 2.3:** The white-matter volume versus the gray-matter volume of the brains of animals spanning a range from Pygmy shrew to Elephant is graphed on log–log graph paper. The solid curve is the best fit of equation (1.18) to the data. (Adapted from [390] with permission.)

most contentious issue in this literature on the brain is the value of the allometry exponent. However, the empirical evidence clearly indicates that $b \approx 1.23$ [390]. He [140] uses scaling arguments and the assumption that the basal metabolic rates (BMRs) of white and gray matter in the brain are equal, to obtain

$$M_W \propto M_G^{6/5}, \tag{2.5}$$

in essential agreement with data. As stated, contained within He's argument is the basal metabolic rate AR, to which we now turn our attention.

### 2.1.2 Physiology

The most studied of the interspecies ARs does not concern relative growth, but is that associating the basal metabolic rate (BMR) $B$ measured in watts, to the total body mass (TBM) measured in kilograms for multiple species, such that [167]

$$B = aM^b. \tag{2.6}$$

The metabolic rate refers to the total utilization of chemical energy for the generation of heat by the body of an animal and is often measured by the oxygen intake during respiration. This remarkable relation covers five decades of data, as depicted in the famous mouse-to-elephant curve depicted in Figure 2.4. Glazier [117] reviews the data for metabolic rates and concludes that 'the 3/4-power scaling law' is not universal either among, or within, animal species. Moreover, the variability in the power-law exponent $b$ is not adequately explained by the then existing theory. He [119] did however, subsequently, develop a theory that did account for much of the variability, which we discuss in Chapter 3.

Heusner [145] adopted geometric scaling arguments to obtain $b = 2/3$ in the AR between BMR and TBM. He argued that the various other values experimentally observed for the power-law index by investigators are a consequence of differing values of the allometric coefficient $a$. He reasoned that two or more data sets with $b = 2/3$, but with different values of $a$, graph as parallel line segments on log–log graph paper, but when the two or more data sets are grouped together and analyzed as a single data set the aggregate is fit by a single line segment with net slope $b > 2/3$. The same argument can be found in a number of other references [129, 267, 288]. Decades earlier, Gould [129] emphasized that an AR only applies to a small range of data to which the parameters are fit, as indicated for the five intraspecies data sets depicted in Figure 2.5. Each data set has the same intraspecies allometry exponent $b = 0.23$, whereas the interspecies allometry exponent is fit by $b = 0.64$. It is evident that the resulting change in the allometry exponent is a consequence of a size dependence of the allometry coefficient, with higher $a$ value being associated with larger sizes.

Heusner [145] concluded that it is the allometry coefficient that remains the central mystery of allometry and not the allometry exponent. We investigate the implications of Heusner's conjecture in Chapter 3, where we explore implications of treating the

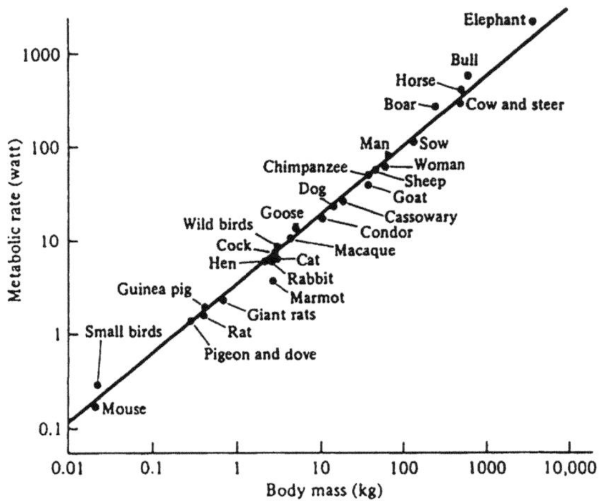


**Figure 2.4:** Mouse to elephant curve. BMR of mammals and birds are plotted versus TBM on log–log graph paper. The solid line segment is the best linear regression of equation (1.18) to the data with a slope very close to 3/4. (From [288] with permission.)

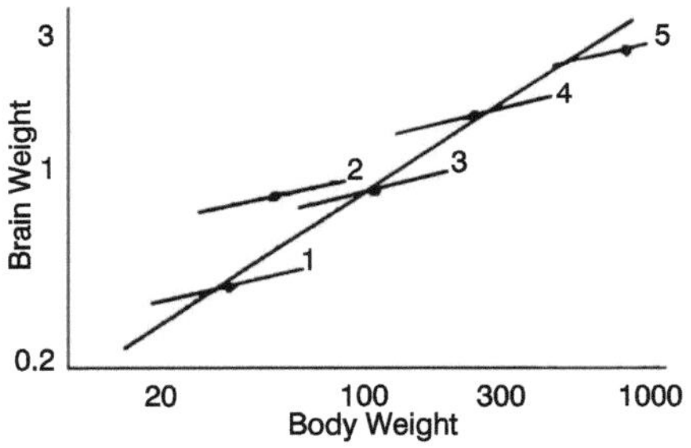


**Figure 2.5:** Intra- and inter-species brain–body curves for five species of insectivores on log–log graph paper with weight in grams are depicted. Each intraspecies AR has an allometry exponent set by its average value of 0.23. The interspecies AR has $b = 0.64$. (Adapted from [129].)

allometry coefficient, as well as the allometry exponent, as random variables. This assumed randomness is no more arbitrary than the more familiar assumption that $a$ and $b$ are empirical constants, as we subsequently demonstrate.

It is useful to list the various forms of physiological allometry given by Lindstedt and Schaeffer [188]: pulmonary and cardiac allometry with $b = -1/4$; renal allometry with $b = -0.85$; liver allometry with $b = -0.85$; pulmonary blood volume with $b = 1.0$; cardiac output with $b = 3/4$; and pulmonary transit times with $b = 1/4$. Their stated intent in compiling the data was to provide the best possible database of normal physiological and anatomical values, primarily for four common mammalian model species: mouse, rat, dog and human. It should not come as a surprise that these cardiac scaling relations have clinical implications [86], but pursuing these applications here would take us too far from our main objective.

### 2.1.3 Time in living networks

The notion of a continuous, linear, unfolding of time was accepted in the experiments of Galileo, explicitly defined by Newton, and not seriously questioned until the revolutionary investigations of Einstein, centuries later. But science does not wait for definitions, it continues forward in exploring phenomena, with or without a clear understanding, confident that such understanding will eventually emerge. The physics of the very fast and the very large does not concern us here, nor does the breath-taking assertion made by a small number of theoretical physicists that time itself may be an illusion [58]. These last concerns bring us back to St Thomas Aquinas, who famously said:

> What, then, is time? If no one asks me, I know what it is. If I wish to explain it to him who asks me, I do not know.

We experience time as a unidirectional flow of conscious experience from the fixed past, through the malleable present, into the unknowable future. But Newton's law of motion predict future behavior, with time moving in both the positive and negative directions without contradiction. The direction of time we humans experience, did not enter into physical theory until complexity was identified and modeled by thermodynamics. The significance of the direction of time was emphasized by Ilya Prigogine in his *Nobel Prize* acceptance speech [257], in which the source of order in non-equilibrium physical/chemical/biological systems was traced to a new type of dynamical state produced by irreversible processes. These new states, called "dissipative structures", are based on the second law of thermodynamics and interrelate time, structure and fluctuations; emphasizing the importance of complexity in non-equilibrium phenomena [258].

The fundamental question addressed herein is whether time for the lumbering elephant is the same as that for a hovering humming bird. Apparently the time shared by the two species is the same, when referenced to a physical clock, but it is not the same, when referenced to their individual physiologies. This difference has lead to such concepts as biological time [383], physiologic time [43], and metabolic time [289], all in an effort to highlight the distinction between time in living networks and that in inanimate networks. The intrinsic time of a biological process was first called biological time by Hill [147], who reasoned that since so many properties of an organism change with size that time itself may scale with TBM.

Lindstedt and Calder [185] developed the concept of biological time further and determined experimentally that biological time, such as species longevity, satisfies an AR with $Y$ being the biological time $\tau$. Lindstedt et al. [187] clarify that biological time $\tau$ is an internal mass-dependent time scale

$$\tau = \alpha M^{\beta}, \tag{2.7}$$

to which the duration of biological events are entrained. They present a partial list of such events that includes breath time, time between heart beats, blood circulation time, and time to reach sexual maturity. In these examples and many others the allometry exponent clusters around the theoretical value $\beta = 1/4$, leading some to the sophisticated speculation that there is a universal allometry exponent in physiological systems [371].

They [186] also point out that attempts to determine specific values of the allometry exponent $b$ presupposes nature to have selected for volume-rate scaling. In a more recent context it has been argued that a fractal network, delivering nutrients to all parts of an organism, is the reason for the existence of physiological ARs. In the latter argument, the metabolic AR is considered to be a consequence of the scaling behavior of the underlying fractal network [365] and therefore the fractal network is thought to be the more fundamental. However, long before the introduction of fractal nutrient networks into the discussion of ARs, Lindstedt and Calder [186] suggested an explanation in which the scaling of biological volume-rates is a consequence of physiologic time scales. For example, the BMR is the energy generated by a biological volume per unit time, such that

$$\frac{\text{volume}}{\text{time}} \propto \frac{M}{M^{\beta}} = M^{b} \tag{2.8}$$

and $\beta \approx 1/4$ implying that $b = 1 - \beta \approx 3/4$ [89, 144, 282]. Consequently, it would not be necessary to hypothesize that specific physiological rates, in this case metabolism, have been selected as isolated phenomena. It would be the physiologic time that makes the metabolic AR inevitable. Lindstedt and Calder [186] emphasize that evolution of change in body size, given by the TBM $M$, affects every aspect of physiology and morphology. Elsewhere [356] we explored the implications of assuming physiological time to be fundamental and will take up the discussion again in Chapter 4.

The term physiologic time was introduced by Brody [43] over a half century ago in recognition of the fact that small animals do not live as long as large animals [310]. A maximum efficiency argument, in support of physiologic time, was advanced by Andresen et al. [7] in which a constant rate of entropy production defined an intrinsic time for the system or 'eigen time'. This constant rate of entropy production is a consequence of the principle of minimum entropy production and the eigen time was identified with physiologic time.

Mordenti [225] provided a graphic representation of the transformation between chronological time and physiologic time as shown in Figure 2.6 for the half-life of expelling drugs from the body. He explained that if various physiologic events in different animals, proceeding at different chronological rates, are instead measured by

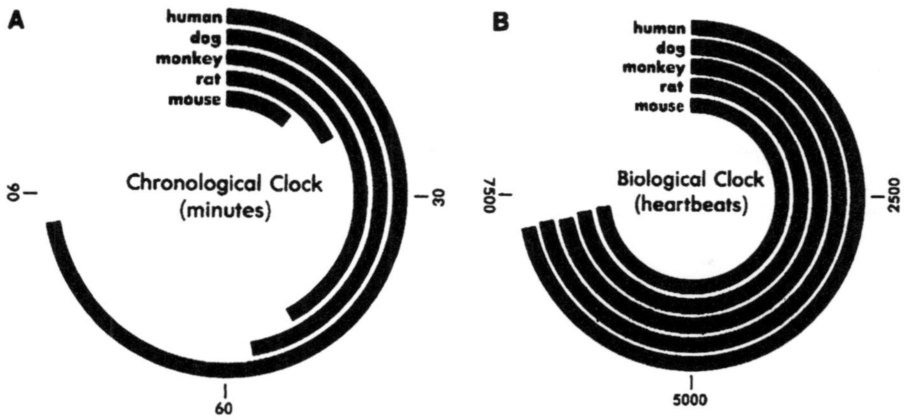


**Figure 2.6:** Preceived differences in the half-life of cefizoxtime in various mammals depend on the reference system used to denote time. (A) When half-lives are reported in minutes, the smaller mammals eliminate 50% of the drug more rapidly than the larger species. (B) When half-lives are reported in the number of heartbeats, all mammals eliminate 50% of the drug in an equivalent time. (Adapted from [225].)

biological clocks, they would then occur in equivalent physiologic time. He expressed this as follows:

> Thus the life-span of an elephant and a mouse is the same when measured with a biologic clock (i.e., heartbeats), although their life spans may vary significantly when measured in years.

Natural scientists have hypothesized that *physiologic time* differs from clock time in that the former changes with the size of the animal [57, 288], whereas the latter does not. On the other hand, the clock time is related to the entropy of a physical system and proceeds irreversibly in the direction of increasing entropy. Herein the rate of entropy increase, together with an assumption regarding the statistics of the spontaneous fluctuations in living systems, are subsequently used to establish the AR for physiologic time. In this respect, time itself becomes a measure of complexity.

The data and data analyses supporting the notion of physiologic time is extensive, see for example, Chapter 6 of Calder [57] and Chapter 12 of Schmidt-Nielsen [288]. In Figure 2.7 we record the average heartrate for sixteen animals [5] as a function of TBM covering six orders of magnitude. The solid line segment is the fit to the data given by

$$R = \frac{205}{M^{0.248}}, \tag{2.9}$$

so that the average heartrate $R$ is consistent with $\beta = 1/4$, given that the physiologic time increases as $1/R$. It is worth noting that we did not fit the logarithm of data using linear regression analysis, but instead used *Mathematica* to fit the nonlinear equation

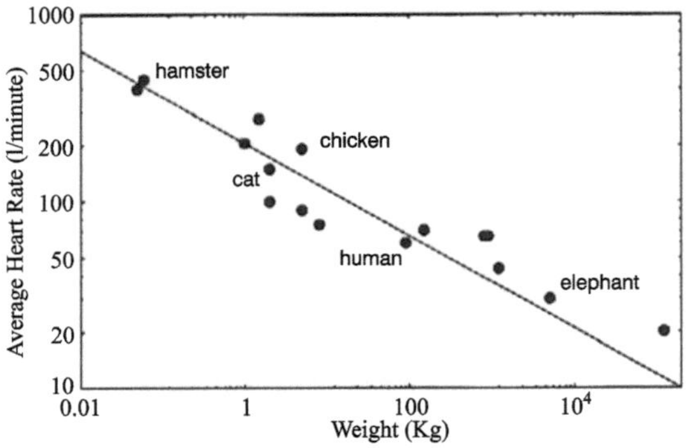


**Figure 2.7:** The average heartrate in beats per minute for 16 animals from the fastest, hamsters, to the slowest, large whales, with humans being in the center of a logarithmic scale. The data were obtained from [5] and the solid line segment is the AR given by equation (2.9) with $r^2 = 0.96$. (From [356] with permission.)

directly with a quality of fit $r^2 = 0.96$. Other, more exhaustive, fits to larger data sets, made by other investigators, may be found in many other places [57, 288], and the results obtained, if not the conclusions drawn, are equivalent.

We [356] mention that the notion of physiologic time has been emphatically rejected by some. Critics, such as Blackstone [38], emphasize that chronological time has no special significance to any organism, making it an invariant standard for interspecies comparisons. Moreover for interspecies rates, neither physiologic time in general, nor size in particular, serve as objective referents in the same way that chronological time does. On the other hand, Heuser [144] argues that animals adapt according to their size, in order to maintain a symbiotic relation with their environment, making physiologic time the more natural measure.

Like most concepts in science the final resolution of the suitability of a concept is its utility in the interpretation of data and how well it interfaces with the larger body of theory. In that regard the speculative nature of physiologic time is no different than the other allometry concepts discussed in this essay.

### 2.1.4 Clearance curves

Another allometry phenomenon is the dependence of drug-dosing range on TBM and is referred to as clearance [151], such as used by Mordenti [225] in the discussion of physiologic time. Zenobiotic clearance is the rate at which any foreign compound not produced by an organism's metabolism is passed from the organism. The application of allometry ideas to pharmacokinetics and to determining human parameters from those in animals is relatively recent [41, 196, 284]. The early studies did not address questions of variability in the allometry parameters and were primarily concerned with whether the allometry exponent more closely tracked the value 2/3 or 3/4, by doing linear regression on log-transformed data [188].

Hu and Hayton [151] addressed the possible impact of statistical variability in the AR parameters on the predicted pharmacokinetic parameter values. They found considerable uncertainty in the value of the allometry exponent, which they fit to a Gaussian distribution with mean value 0.74 as depicted in Figure 2.8. Even though they could not determine whether the variability in the allometry exponent was due to experimental error, or to biological mechanisms, they did find that there was no systematic deviation from the AR. However it appears that whether $b = 3/4$ or $2/3$ depends on which of the major elimination pathways is used, metabolism for the 3/4 value and renal excretion for the 2/3 value.

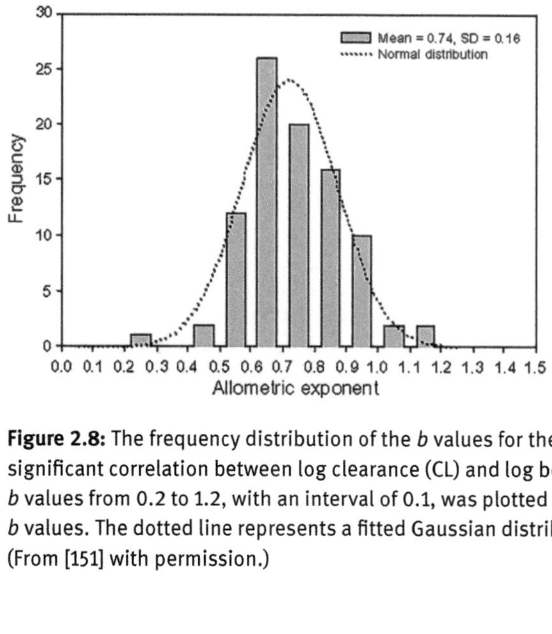


**Figure 2.8:** The frequency distribution of the $b$ values for the 91 xenobiotics that showed statistically significant correlation between log clearance (CL) and log body weight (BW). The frequency of the $b$ values from 0.2 to 1.2, with an interval of 0.1, was plotted against the midpoint of each interval of $b$ values. The dotted line represents a fitted Gaussian distribution curve. SD = standard deviation. (From [151] with permission.)

### 2.1.5 Botany

Niklas [233] shows in Figure 2.9 an impressive statistical trend spanning twenty orders of magnitude in the mass of aquatic and terrestrial nonvascular, as well as, vascular plant species. The annual growth in plant body biomass $G_T$ (net annual gain in dry mass per individual) and $M_T$ (total dry mass per individual) are related by the empirical AR:

$$G_T \propto M_T^{3/4}. \tag{2.10}$$

Figure 2.9 shows linear regression on logarithmically transformed data

$$\log G_T = \log a + 0.75 \log M_T, \tag{2.11}$$

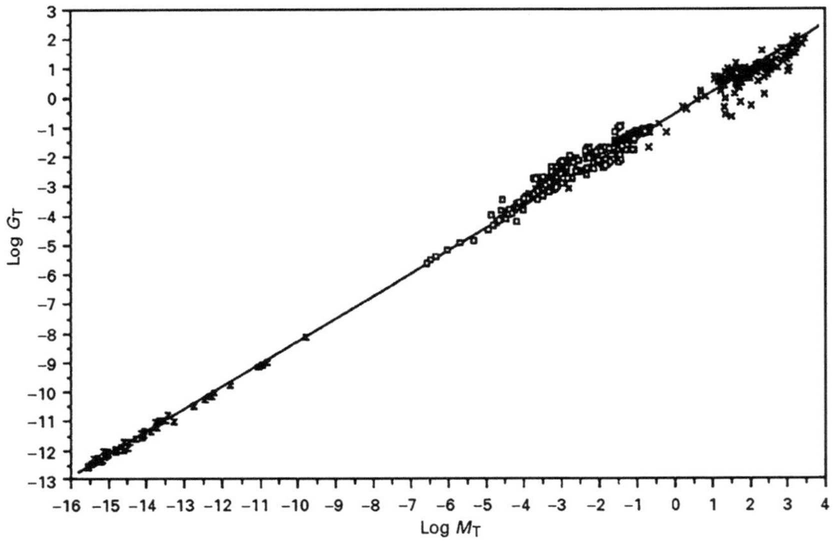


**Figure 2.9:** Log–log bivariate plot of total annual growth rate in dry body mass per individual $G_T$ versus TBM. Line segment denotes reduced major axis regression curves for the entire data set. (Adapted from [233] with permission.)

and the parameter $a$ is determined by the intercept with the vertical axis. In the data analyses recorded here, the terms weight, mass and volume are used almost interchangeably, as measures of size. The allometry exponent is 3/4 for the data in Figure 2.9, but empirically differs from this value when the data sets are graphed individually; recall the discussion surrounding Figure 2.5. The allometry coefficients of the separate data sets may vary as a function of habitat as well. The agreement between the biomass data and the AR with exponent 3/4 is very suggestive, but it must be viewed critically because of methodological limitations.

Reich et al. [265] analyzed data for approximately 500 observations of 43 perennial plant species of coupled measurements of whole-plant dry mass and $G_T$ from four separate studies. Collectively, the observations span five of the approximately 12 orders of magnitude of size in vascular plants [95]. The result of each experiment yielded an isometric scaling of $b \approx 1$ and not $b = 3/4$ as did the scaling of $G_T$ to TBM for whole plants. Consequently, even when data look as appealing as they do in Figure 2.9, things are not always what they seem.

### 2.1.6 Information transfer and Rent's Rule

There are literally dozens of physiologic ARs for physiologic time $\tau$, relative to clock time $t$, that increases with increasing body size as given by equation (2.7) [186] and describes chemical processes, from the turnover time for glucose with $\beta = 1/4$ [16], to the life span of various animals in captivity with $\beta = 0.20$ [277]. Schmidt-Nielsen [288] explains how a variety of physiologic time scales, such as the length of heartbeat and respiration intervals, all scale with body size and from that deduce a number of interesting relations. It is only recently, however, that Hempleman et al. [142] hypothesized a mechanism to explain how information, about the size of an organism, is communicated to the organs within the organism. Their hypothesis involved matching the neural spike code to body size, in order to convey this information.

Hempleman et al. [142] suggest that mass-dependent scaling of neural coding may be necessary for preserving information transmission efficiency, with decreasing body size. They point out that action potential spike trains are the mechanisms for long distance information transmission in the nervous system. They go on to say that neural information may be 'rate coded', with average spike rate over a time period, encoding stimulus intensity or 'time coded', with the occurrence of a single spike encoding the occurrence of a rapid stimulus transition. The hypothesis is that some phasic physiological traits are sufficiently slow in large animals to be neural rate coded, but are rapid enough in small animals to require neural time coding. These traits include such activities as breathing rates that scale with $\beta = -1/4$.

They tested for this allometry scaling of neural coding by measuring action potential spike trains from sensory neurons that detect lung $CO_2$ oscillations, linked to breathing rate in birds ranging in body mass from 0.045 kg to 5.23 kg. While it is well known that spike rate codes occur in the sensing of low frequency signals and spike timing codes occur in the sensing of high frequency signals, their experiment was the first designed to test the transition between these two coding schemes in a single sensory network, due to variation of body mass. The results of their experiments on breathing rate was an allometry exponent in the interval $-0.26 \leq \beta \leq -0.23$ and, although taken on a small number of birds, their results do suggest a preservation of information transmission rates for high frequency signals in intrapulmonary chemoreceptors and perhaps other sensory neurons as well. The implications of these experiments strongly suggest the need to continue such investigations.

On the more theoretical side, Moses et al. [227] apply the scaling ideas of metabolic allometry, also developed by West et al. [365] , to explain the AR depicted in Figure 2.4, to information networks consisting of microprocessors to form a network. Moses et al. [227] use an argument involving fractals to construct a two-dimensional hierarchal self-similar branching network, an H-tree motif that Mandelbrot [201] originally used in his discussion of the space filling behavior of the human lung. They show that this branching network of microprocessors have a striking similarity to such networks in organisms, even though the latter has evolved by natural selection and the former imposed by engineering design.

Along this same line E. F. Rent, while a scientist/engineer at IBM in the 1960s, wrote a number of internal memos (unpublished) relating the number of pins at the boundaries of an integrated circuit ($X$), to the number of internal components ($Y$), such as logic gates. In this way he obtained an AR with $\beta < 1.0$. Rent's Rule, or the Rent AR, has historically been used by engineers to estimate power dissipation in interconnections and for the placement of components in very large scale integrated (VLSI) circuit design. An excellent review of Rent's Rule and the placement models derived from it is given by Christie [70], wherein the power-law form of the AR is shown to predict the number of terminals required by a collection of gates to efficiently communicate with the rest of the circuit and to be a consequence of statistically homogeneous circuit topology and gate placement.

More recently, the Rent AR has been used to model information processing networks within the human brain [24], where the mass of gray and white matter are shown to satisfy an AR as first noted by Schlenska [287]. Beiu and Ibrahim [31] suggested that the allometry exponent for gray and white matter between species is identical to the Rent exponent within a species and this was supported using magnetic resonance imaging (MRI) data by Bassett et al. [24], as depicted in Figure 2.10.

Bassett et al. [24] maintain that the exponent of the interspecies brain volume AR is simply related to the Rent exponent of mammalian cerebral connectivity, using data from [55]. Lines fitted to the data show the AR scaling predicted by Rent's exponent, estimated for neuroimaging data on a single species (Homo sapiens), using magnetic resonance imaging (MRI, $\beta \approx 0.828$) and DSI ($\beta \approx 0.782$).

## 2.2 Physical networks

Some of the oldest ARs involve physical networks, or more specifically geophysical networks. The skeptic need only return to da Vinci's scaling relation to see the geophysical application. In his notebooks da Vinci explains the meaning of his scaling equation, not only in the context of relating tree trunks to subsequent branches, but to the branchings of rivers and tributaries, as well. Long before the conservation of energy and the continuity of fluid flow were envisioned by scientists, the enigmatic Italian painter, sculptor, military engineer and anatomist, sketched the basics for the scaling of hydrodynamic networks.

### 2.2.1 Geology and geomorphology

The fractal patterns of rivers and their branches are dramatically evident from satellite pictures, such as those of the Amazon River shown in Figure 2.11. It is quite remarkable that such patterns were evident to geologists and geophysicists long before such photographic images became available. The patterns were so well known that they were given a mathematical description. For example, Horton's Law of river numbers is an empirical regularity observed in the topology of river networks [149]. As observed by Scheidegger [285] the number of river segments in successive order form a geometrical

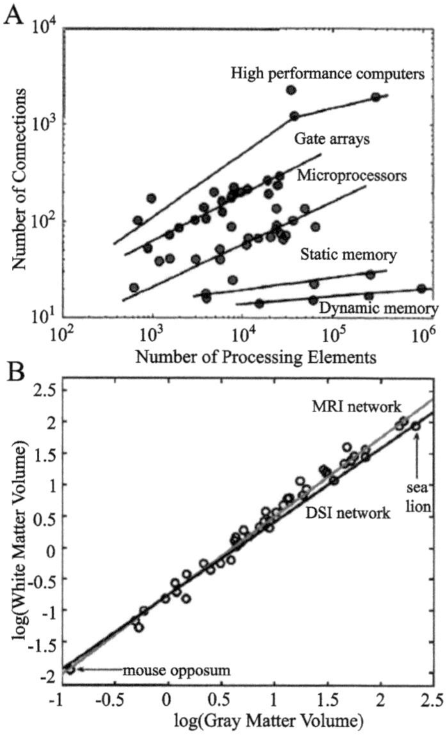


**Figure 2.10:** AR scaling of VLSI circuits and mammalian brains are depicted. (A) In computer circuits, the number of connections at the boundary of a chip scales following Rent's AR, with the number of processing elements; the exponent is greater for high performance computers than for simpler dynamic memory circuits. (B) In the cerebral hemispheres of mammalian brains, there is an AR scaling relationship, between white matter volume (related to connectivity to elements) and gray matter volume (related to number of processing elements). Errors in the fits are smaller than the line width. (Adapted from [24] with permission.)

sequence such that:

$$n_k \equiv \text{number of rivers with } k \text{ tributaries} \propto R_b^{1-k}. \tag{2.12}$$

The bifurcation parameter $R_b$ is the constant ratio between successive numbers of river networks, known as Horton's Law of stream numbers

$$n_k/n_{k+1} = R_b. \tag{2.13}$$

This parameter has an empirical value between 4.1 and 4.7 in natural river networks [241, 242]. By way of contrast, the random model of river networks predicts a value of 4. Note that the system of counting begins at the smallest tributaries, which are the most numerous $R_b > 1$, and these feed into larger tributaries that are fewer in number.

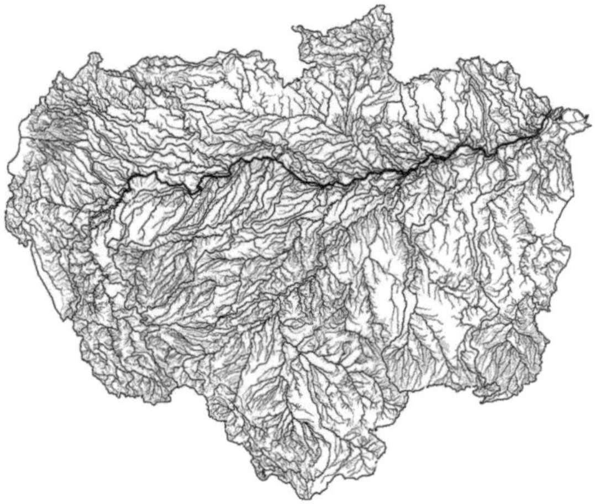

**Figure 2.11:** The Amazon river and its tributaries are highlighted from a satellite image.

Dodds and Rothman [88] point out that universality arises when the qualitative character of a network is sufficient to quantify its essential features, such as the exponents that characterize scaling laws. They go on to say that scaling and universality have found application in the geometry of river networks and the statistical structure of topography within geomorphology. They maintain that the source of scaling in river networks and whether or not such scaling belongs to a single universality class is not yet known. They do provide a critical analysis of the Hack Law, see also, Rodriguez-Iturbe and Rinaldo [275].

### 2.2.2 Hydrology

Hack's Law is a hydrologic AR having to do with the drainage basins of rivers. Hack [135] developed an empirical relation, between the mainstream length of a river network and the drainage basin area at the closure of the river. The law is given by,

$$\text{main stream length} \propto (\text{area})^{h}, \tag{2.14}$$

where $h$ is the Hack exponent with the typical empirical value $h \approx 0.57$, see figure 2.12. Hack asserted, without proof, that river networks are not self-similar.

Mandelbrot [201] relates the Hack exponent to the fractal dimension of the river network and questions the earlier interpretation of the equation. Feder [99] observed that defining a fractal dimension for river networks was obscure and required further study. A modern version of this discussion in terms of hydrologic allometry is given by Rinaldo et al. [272] who point out that optimal channel networks yield $h = 0.57 \pm 0.02$ suggesting that *feasible optimally* [270] implies:

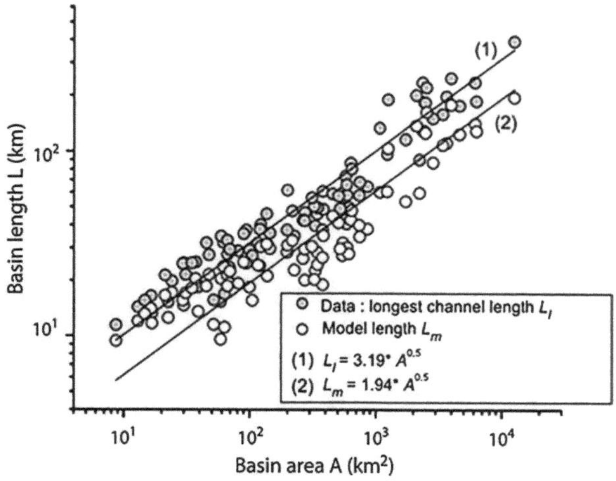


**Figure 2.12:** Hack's Law for the studied basins with longest channel length (gray) and modeled basin length (white). While the Hack exponent of longest channel length is subject to debate, the coefficient in the law of basin length corresponds to the average shape of the studied basins. However, all basins are different, and the 0.5 self-similarity exponent in Hack's Law is rather a result of averaging inside the restricted range of possible basin shapes. (From [62] with permission.)

> ...Hack's law, one of the most common empirical relationships found in the river basin, is naturally derived from optimality principles only in connection to the nature of the stationary states achieved. Dynamically accessible stationary states, found by imperfect search procedures seemingly relatable to nature's myopic search for a stable niche within a complex fitness landscape, show a striking resemblance to real basins. On the contrary, globally optimal configurations, obtained by screening procedures that allow unphysical freedom to the search process, do not elongate. We thus conclude that feasible optimality is likely to be the product of the self-organized dynamics of fluvial networks and that Hack's law, whose ubiquity is widely acknowledged reinforces the conclusions of previous work, suggesting that frustrated optimality of complex dissipative systems with many degrees of freedom might be the dynamic reason for scale-free growth and form in nature.

Another viable model is given by Sagar and Tein [278] that is geomorphology realistic, giving rise to general ARs in terms of river basin areas, as well as parallel and perpendicular channel lengths. Maritan et al. [206] consider an analogy with the metabolic AR using

$$M \propto B^{\alpha},$$

that is, with $\alpha = 1/b$ we obtain $\alpha = 1 + h$, with the limiting values $\alpha = 3/2$ and $h = 1/2$, in the case of geometric self-similarity. Geometric self-similarity is the preservation of the river's shape as the basin increases in area. The observed values lie in the range $1.5 \le \alpha \le 1.6$ and the scatter of individual curves (analog of intraspecies data) is remarkably small. These values suggest that the branching nature of rivers is fractal, that is, $\alpha > 3/2$ in most cases. The ensemble average of the Hack exponents from different basins extend over 11 orders of magnitude and is indistinguishable from $h = 1/2$ [222]. Maritan et al. [206] conclude that like the interspecies metabolic rate, the slope of the intraspecies $h$'s are washed out in the ensemble average, resulting in the value $h = 1/2$.

## 2.3 Natural History

Natural History embraces the study, description and classification of the growth and development of natural phenomena. The focus of investigation includes, such important contemporary areas as, ecology and paleontology, parts of which rely heavily on ARs and scaling.

### 2.3.1 Wing spans

One of the more unexpected ARs arises in the allometry of wings. Wings are an example of macro-evolutionary adaptation of animals to their environment, where the aerodynamic forces controlling flight, such as drag and lift, select for physical structures that enhance fitness over evolutionary time scales. Such adaptation results in the various forms of flight observed, for example: flapping forward flight; gliding flight at fixed angle and zero stroke frequency; bounding flight; soaring flight of the albatross; hovering flight of insects and humming-birds and the clap and fling movement of chalcid wasps [226].

The original explanation for wing allometry was in terms of geometric scaling, as had been done for other biological ARs by McMahon [210], which is discussed subsequently. These geometric scaling arguments predict an allometry exponent 1/3 for any length scale and 2/3 for any area, just as predicted for heat balance in the metabolic AR. However, the observed value of this parameter for wing span was $b = 0.394$ as shown in Figure 2.13 for a collection of 70 Chilean birds, and not the 1/3 from geometric scaling. Other empirical allometry exponents include that for a flight surface $b = 0.664$ in agreement with the 2/3 from geometric scaling and for single-beat frequency $b = -0.382$, rather than $-1/3$. In each case the correlation coefficient $r^2$ was near unity [226].

Morgado et al. [226] presented additional data for different orders of birds, classified according to the aerodynamic criterion of wing loading, that is, the TBM supported per unit wing area, ranging in size from (10 g, 150 mm) to (3.6 kg, 1.62 m). A typical value of the allometry exponent obtained for one such aerodynamic bird model was $b = 0.427 \pm 0.01$, which is in essential agreement with earlier studies [264]. They also confirmed the wing span AR of hummingbirds [131] to have the allometry exponent $b = 0.53 \pm 0.03$. They conclude by pointing out that a spectrum of allometry exponents have been obtained for wing spans as a function of TBM, ranging from 0.30 for bats to

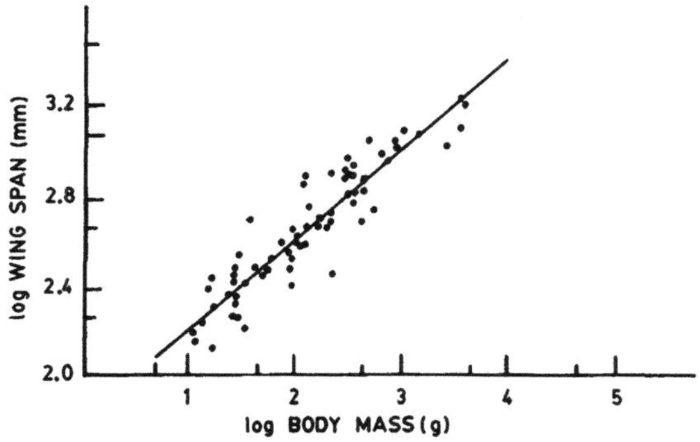


**Figure 2.13:** Log–log plot of wing span (mm) as a function of body mass (g) in 70 Chilean birds. The mean value of the allometry exponent is $b = 0.394 \pm 0.02$ and the quality of fit of the data to the empirical AR is $r^2 = 0.839$. (Adapted from [226] with permission.)

0.53 for hummingbirds and approximately 0.40 for three bird models. This variability in allometry exponent is explained by them as being due to the fact that wing spans and wing areas can change markedly during each wing stroke. But even with this variability the wing span AR persists.

### 2.3.2 Ecology

Ecology is the scientific study of the distribution, abundance and relations of organisms and their interactions with the environment. Such living networks include both plant and animal populations and communities, along with the network of relations among organisms of different scales of organization. Of concern to us here are the species traits determined by body size and how these, in turn, may affect food web stability. Woodward et al. [384], along with many others, point out that the largest metazoans, for example, whales ($10^8$ grams) and giant sequoias, weigh over 21 orders of magnitude more than the smallest microbes ($10^{-13}$ grams) [158, 343]. They go on to stress the considerable variation in TBM among members of the same food web.

Hatton et al. [137] address the traditional linking of species through the mechanisms of the prey-predator relations within food webs [125]. They emphasize that the AR pattern is only observed over large heterogeneous populations and emerges uniquely at the ecosystem level. They observe the pattern depicted in Figure 2.14, but freely admit that they do not know why the pattern occurs, given that it is not predicted by current theoretical models and is unexpectedly independent of lower-level structure. They remark further that the same pattern depicted in this figure, where the data is drawn from prey-predator biomass in African savanna, systematically recurs in forests, lakes and oceans, as well as, in other grasslands.

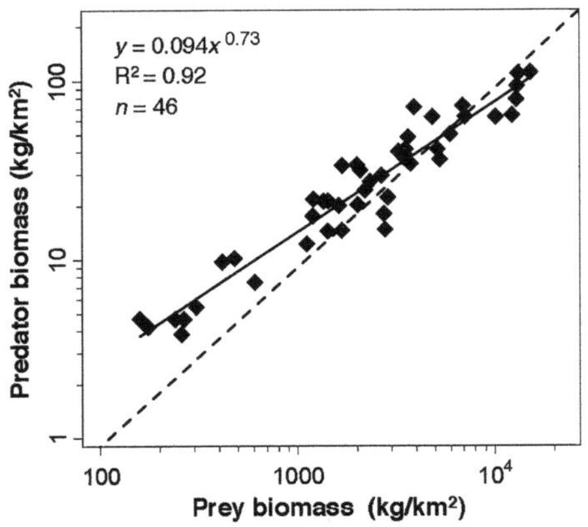


**Figure 2.14:** African prey-predator communities exhibit ecosystem structures. Predators include lion, hyena, and other large carnivores (20 to 140 kg), which compete for large herbivores prey from dik-dik to buffalo (5 to 500 kg). Each point is a protected area. This near 3/4 scaling law is found to recur across ecosystems globally. (Adapted from [137] private communication.)

The significance of body size has been systematically studied in ecology [50, 71, 158]. Identifying $Y$ with species abundance and $X$ with TBM, there is, in fact, an AR between the species at the base of a food web and the largest predator at the top [71]. We note that species-area power functions have a vital history in ecology [254, 379], even though the domain of sizes over which the power law appears valid is controversial [48, 378].

Woodward et al. [384] emphasize that AR has been used to explain the observed relations between body size and a variety of functions, including home range size, nutrient cycling rates, numerical abundance, as well as, biomass production. They speculate that body size may capture a suite of covarying species traits into a single dimension, without the necessity of having to observe the traits directly.

Brown et al. [49] discuss the universality of the documented ARs in plants, animals and microbes; to terrestrial, marine and freshwater habitats; and to human-dominated as well as 'natural' ecosystems. They emphasize that the observed self-similarity is a consequence of a few basic physical, biological and mathematical principles; one of the most fundamental being the extreme variability of the data. The variety of distributions of allometry coefficients and exponents are discussed both phenomenologically and theoretically in subsequent chapters.

### 2.3.3 Zoology and acoustics

Mice squeak, birds chirp and elephants trumpet due to scaling. Fitch [106] discusses the relationship between an organism's body size and acoustic characterization of its vocalization, under the rubric of acoustic allometry. Data indicate an AR between palate length (the skeletal proxy for vocal tract length) and TBM for a variety of mammalian species. He shows that the interspecies allometry exponent attains the geometric value of 3 in the regression of skull length and TBM, whereas the intraspecies allometry exponent varies a great deal. The significant variability in the intraspecies allometry exponent suggest taxon-specific factors influencing the AR [207, 295].

Fitch [106] gives the parsimonious interpretation that the variability in the intraspecies allometry exponent could be the result of each species adopting allometry scaling, during growth as postulated by Huxley, with a different proportionality factor for each species. On the other hand, the interspecies allometry exponent could result from the common geometric constraints across species, due to the wide range of body sizes. He concludes that the AR, between vocal tract dimensions and body, could provide accurate information about a vocalizer's size in many mammals.

### 2.3.4 Paleontology

Pilbeam and Gould [249] provide reasons as to why body size has played such a significant role in biological macroevolution. The first is the statistical generalization known as Cope's Law, which states that population lineages increase in body size over evolutionary time scales [74, 261], that is, the body size of a species is an indication of how long it has survived on geological time scales. A second reason is the one mentioned earlier, Galileo's observation that as organisms increase in size, they must change shape in order to function in the same way as do their smaller prototypes.

One quantitative measure of evolution is the development of the brain in mammals at various stages of evolution. Jerison [155] showed that the brain–body AR is satisfied by mammals, with an exponent that is statistically indistinguishable from 2/3. He suggested that $a$ may be an appropriate measure of brain evolution in mammals as a class, following the proposal of Dubois [90] that a quantitative measure of cephalization (development of a head by the concentration of feeding and sensory organs and nervous tissue at the anterior end) in contemporary mammals be based on the ratio:

$$a = \text{brain weight}/(\text{body weight})^b. \tag{2.15}$$

These hypotheses were directly tested by Jerison [155], using endocranial volumes and body volumes for fossil mammals at early and intermediate evolutionary stages. The data did in fact support the hypothesis.

White and Gould [372] emphasize in their review of the meaning of the allometry coefficient $a$ that it had generated a large and inconclusive literature. Reiss [267] notes that if brain mass is regressed on TBM, across individuals in a species, the slopes are shallower than of regressions calculated across mean values for different species, within a single family (genus). This argument had previously been presented by Gould [129], who emphasized the importance of the allometry coefficient in the geometric similarity of allometric growth. His interpretation of the allometry coefficient is at odds with the belief, of the majority of the scientific community, that the allometry coefficient is independent of body size. This latter view is also contradicted by the data analysis presented in Chapter 3, where the covariation of the allometry parameters is discussed.

Allometry has been used by Alberch et al. [3], as the first step in creating a unifying theory in developmental biology and evolutionary ecology, in their study of morphological evolution. They demonstrate how their proposed formalism relate changes in size and shape, during ontogeny and phylogeny.

## 2.4 Sociology

In the allometry context, the aspect of sociological interest are the patterns that emerge in data that depend on the size of the social group. In large urban centers, size dependencies are found in a city's physical structure [27], as well as in the city's functionality, such as in the behavior of people [217]. Among the most compelling aspects of urban life, for example, income level and innovation are shown to be allometry phenomena.

### 2.4.1 Effect of crowding

Farr's Law [51] demonstrates the change in ARs making the transition from organismic to environmental allometry. Farr collected data on the number of patients committed, because of their mental condition, as well as, data on their mortality, from a variety of asylums in 1830s England [98]. From these humble beginnings, he was able to summarize the "evil effect of crowding" into a relation between mortality rate $R$ and average population density $\rho$ [150]:

$$R = a\rho^b. \tag{2.16}$$

Here we see that the size measure used in the metabolic AR, the mass, is replaced with a measure of community structure, the population density.

Farr's Law stipulates that the mortality rate increases with average population density since $b > 0$, and the more people packed into a given area, the greater the likelihood of any one of them dying, on average. The ARs that capture life histories in ecology and sociology are often expressed in terms of numbers of animals and areas in addition to body mass. Calder [57] points out that size and time seem to be the principle characteristics of life history and ecology.

The factors necessary for the formation of social groups, within a restricted geographic area, are not completely understood, but certain ARs help clarify some of them. The average population density $\rho$ of herbivorous mammals has been determined to be related to their average mass $M$ by the Damuth Law [79]:

$$\rho^{-1} = cM^{0.75}, \tag{2.17}$$

where the animals freely roam on a 'home range' of given area

$$A_{\mathrm{hr}} = c'M^{1.02}. \tag{2.18}$$

A similar relation exists for carnivores [221]. Consequently, as explained by Calder [57], herbivorous mammals, above a certain size, range over an area greater than their per capita share of the local habitat. The degree of overlap was empirically determined to be [80]

$$A_{\mathrm{hr}}\rho = c''M^{0.34}, \tag{2.19}$$

where the empirical exponent $0.34 \pm 0.11$ is statistically indistinguishable from that obtained by combining the separate exponents for the population density and area, that being, $1.02 - 0.75 = 0.27$. Calder conjectures that the greater the product $A_{\mathrm{hr}}\rho$, the greater the intensity of competition and the greater the desirability of social networks that contribute to mutual tolerance within these groups. Makarieva et al. [199] argue that animal home ranges represent biological footprints of the undisturbed state of an econetwork, however the population density adapts to disturbances in the econetwork. Consequently, the deviation of the home range-population product from isometry, that is, the deviation of $b$ from one, reflects the degree of econetwork disturbance.

The AR between maximum abundance and average body size for terrestrial plants

$$N \propto M^{-b}, \tag{2.20}$$

was extended by Belgrano et al. [32] to the maximum population densities of marine phytoplankton with $b = 3/4$. They draw the implication that maximum plant abundance is constrained by rates of energy supply, in both terrestrial and marine networks, as dictated by a common AR. Earlier investigators found $b > 3/4$ [77, 91].

### 2.4.2 Urban allometry

There appear to be certain universal rules for how humans allocate or partition land. Fialkowski and Bitner [105] studied the morphology of parcel patterns, constructed by humans in both urban and rural regions. For parcels of size $\mathcal{A}$ the PDF (histograms) $p(\mathcal{A})$ was determined to have three morphological types: the city core was captured by an IPL:

$$p(\mathcal{A}) = a\mathcal{A}^{-b}, \tag{2.21}$$

with $b \approx 2$. The core is surrounded by a ring of suburban area, with a log-normal parcel PDF. Finally, rural areas again have the IPL PDF, but with $b \approx 1$. The majority of the 33 cities used in their study were in Australia and the United States. Fialkowski and Bitner conclude that the morphology of the parcel pattern conforms to those observed in natural phenomena and enable an unambiguous classification of land use as city core, suburbs, or rural.

Batty et al. [28, 29] examine urban spatial structure in large cities through the distribution of buildings in terms of their volume, height and area, while maintaining that there is no well worked out theory of urban allometry. As they point out, the allometry hypothesis suggests the existence of critical ratios between geometric attributes that are fixed by the functioning elements, just as in living organisms. An example they use is the dependence of natural light on the surface area of a building, so that to maintain a given ratio of natural light to building volume, the shape of the building must change with increasing size. Consequently, the volume is not given by the surface area raised to the 3/2 power, but is found empirically to have an allometry index $b \approx 1.3$. They interpret this to mean that the volume does not increase as rapidly, with increasing surface area, as it would for strict geometric scaling, or strict rationality on the part of the builder, which is to say, the aesthetic always has an element of the irrational imbedded within it. A number of such ARs are determined to interrelate the volume, area, height and perimeter of buildings, indicating the strong influence of allometry on human design and sense of what is attractive.

Bettencourt et al. [35] point out that cities have, at all times and in all places, throughout history, produced extremes in human activity, generating creativity, wealth, as well as crime. Moreover, for the first time in history the majority of people are now living in cities and so the need to understand the functioning of these vast social organization is more important than it has ever been. The AR for wealth and innovation in urban centers is concave, with an allometry exponent for the population greater than one ($b > 1$), whereas the ARs accounting for infrastructure are convex ($b < 1$). The convex urban ARs share the economy of scale that is enjoyed by biological networks, since

$$Y/X \propto 1/X^{1-b} \tag{2.22}$$

decreases with the network size. As they also point out, economy of scale facilitates the optimized delivery of social services, such as healthcare and education. The interpretation of cities in terms of biological metaphors is intellectually useful, but the existence of ARs for social organizations is certainly more than a metaphor, as shown for the case of wages in Figure 2.15. Note also that the nearly ubiquitous 1/4 that appears in all manner of theory to explain biological allometry exponents, is not evident for urban ARs.

They [35] go on to contrast the convex, with the concave, urban ARs that focus on the growth of occupations oriented toward innovation creation, that being, rates of invention, measured by new patents and employment in creative sectors and wealth

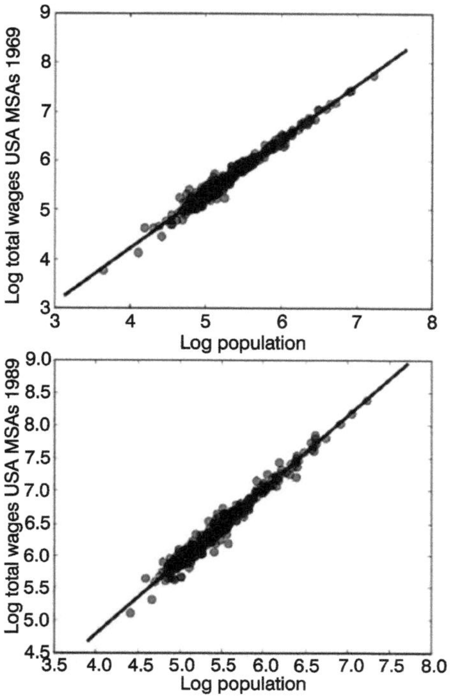


**Figure 2.15:** The consistency of scaling relations for socioeconomic outputs over 40 years is depicted. Scaling of total wages (in thousands of US$) in US Metropolitan Statistical Areas in 1969, $b = 1.11$ (95% confidence interval [1.09, 1.13], $r^2 = 0.95$) upper; in 2008, $b = 1.10$ (95% confidence interval [1.08, 1.13], $r^2 = 0.93$) lower. The exponents remain within each other's 95% confidence intervals, as emphasized by Bettencourt [37], even as population and baseline wages, determined by the allometry coefficient, grew considerably and the economic base of cities changed profoundly, undergoing several periods of recession and expansion. (Adapted from [37] with permission.)

creation, that being, wages, income, growth of domestic product, and bank deposits. Of particular interest is their discussion of the scaling of rates of resource consumption, with city size in direct correspondence to our earlier comments on physiologic time. They emphasize the concave situation for processes driven by innovation and wealth creation, having $b > 1$, manifest an increasing pace of urban life within larger cities, as first observed by Milgrim [217], and which they quantitatively confirm increasing urban crime rates, the spread of infectious diseases and pedestrian walking for increasing city size.

Bettencourt [37] shows how the urban AR with $b > 1$ is a property of many cities, across the Americas, Asia, and Europe. Figure 2.15 shows the extraordinary consistency of superlinear scaling of wages in US Metropolitan Statistical Areas, over the last 40 years, see also [36]. Throughout this period the allometry exponents stayed remarkably consistent being on the order of 1.1, as depicted in Figure 2.15. Bettencourt

[37] emphasizes that wages, along with the other data, confirms the prediction that the corresponding allometry exponents are independent of time and the levels of socioeconomic development, which vary considerably over this period.

## 2.5 Summary

In this chapter we mentioned over thirty different discipline-specific ARs. Some empirical laws are part of the same discipline; such as those of Hack and Horton, but refer to different aspects of geomorphology; or the many 1/4-power scaling laws in physiology; and the multiple forms of biological time, whether as a heartrate or the frequency of beating wings. The most significant aspect of the presentation is the fact that ARs from the disciplines of biology, botany, computer science ecology, geology, hydrology, paleontology, physiology, sociology, and zoology, are all given without a single mechanism in common. What can we make of this?

It is clear that the full spectrum of ARs is independent of any discipline-specific mechanism, and perhaps is independent of any finite collection of such mechanisms. But the mechanisms do share a common feature; their tie to the underlying complexity of the phenomenon being addressed within the discipline. These examples support the hypothesis made in Chapter 1 that the empirical AR given by equation (1.19) is a consequence of the imbalance between the complexity associated with system functionality and that associated with the system size. The allometry/information hypothesis is supported by the fact that within a complex network, composed of two or more interacting sub-networks, such as mortality rate and population density, or the white and gray matter of the brain, to name two, the flow of information is driven by the complexity gradient between the sub-networks. Information flows from the sub-network with the greater to that with the lesser complexity, thus, producing an information force that entails an AR within the complex host network.

# 3 Statistics, scaling and simulation

In this chapter statistical assumptions are introduced to analyze functionality and size measurements from various systems, as well as, to identify patterns in the data (empirical laws) and to develop alternative methods of statistical analysis. Warton et al. [327] point out that fitting a function to a bivariate data set is not a simple task and the AR literature has more than its share of, sometimes acrimonious, debates over what constitutes proper methodology. Sir Julian Huxley, adopted the statistical approach of linear regression to fitting logarithmically transformed data sets, to determine the empirical allometry coefficients $a$ and exponents $b$. Sophisticated statistical techniques, such as *principle component analysis*, were not available to Sir Julian and, although they can be found in the modern AR literature, least-square linear regression still appears to be the method of choice to the modern investigator [119, 282].

Measures of complex phenomena always contain uncertainty, either due to internal dynamics or environmental influences, resulting in data sets, ostensibly measurements of the same quantity, that vary randomly from measurement to measurement. Thus, the size $X(t)$ and functionality $Y(t)$ of a given complex network are stochastic variables, and by implication $Y(X)$ is a random function of its argument, as well. Consequently, any deterministic algebraic equations, relating size and functionality, never unambiguously represent what is actually being measured. At best, the ARs denote relationships between average quantities. Subsequently, we examine the appropriate mathematical modeling, shifting our focus from dynamic stochastic variables to the associated dynamic PDFs. This is necessary because the AR encapsulates a network property that is only manifest in a PDF, or more accurately, in the information contained in the PDF.

We examine the phenomenology of the random data from measurements of various properties of allometry networks that determine empirical ARs. In particular, we focus on how allometry coefficients and exponents are interpreted, given that the data on which they are based fluctuate to such a large extent. Part of the reason for taking this approach is that a significant number of scientists adopt the viewpoint that erratic fluctuations indicate a lack of control and/or ignorance about what is being measured. Here we adopt the more sympathetic view that network variables are inherently random, because of their complex dynamics, in which case their statistics provide information about the fundamental nature of that complexity, when properly analyzed.

As we mentioned, the data pairs $(X_{ij}, Y_{ij})$ display a set of randomly fluctuating points in a two-dimensional space, where $j = 1, \ldots, N$ labels the $N$ measurements for each species $i$. Another set of randomly fluctuating data points is given by the average values $(\langle X_i \rangle, \langle Y_i \rangle)$, since there is nothing to indicate that the average values change in a predictable way from species to species. But such data spaces are not unique, when an AR is assumed to model the data. There are two equally reasonable ways to inter-

DOI 10.1515/9783110535136-003

pret the fluctuating average data. The traditional interpretation is that the allometry coefficient and exponent are fitting parameters and the functionality and size variables are random. Given an empirical AR, one could, just as easily, interpret $X$ and $Y$ as fitted quantities and the allometry parameters as random. In this latter approach the two-dimensional parameter space $(a, b)$ contains the fluctuations. We find it useful to adopt this latter perspective here and determine the PDFs for the AR parameters from histograms of the fluctuating quantities.

## 3.1 Interpreting fluctuations

All complex dynamic networks have fluctuations, either due to intrinsic nonlinear dynamics, producing chaos [190, 191, 236], or due to coupling of the network variables to an infinite dimensional, albeit unknown, environment [184], or both. These random fluctuations have nothing to do with measurement error and the LFE. Consequently, it is necessary to understand how statistical uncertainty may be included in modeling the dynamics of allometry variables. Kaitaniemi [163] pointed out that the potential information content of the allometry coefficient has been largely neglected, an observation also made by Glazier [119], among others. Kaitaniemi examined the different ways the allometry coefficient may vary for different sources of random fluctuations. Here we follow a similar strategy, but we use empirical rather than computer-generated random data to make the same point.

The normal or Gauss PDF suggests that the statistical variations between the variables in an AR may be additive, leading some scientists [116, 200, 300] to propose the perturbed form of the schematic AR:

$$Y = \overline{a}X^{\overline{b}} + \eta, \tag{3.1}$$

where $\eta$ depicts random fluctuations and the overbars on $\overline{a}$ and $\overline{b}$ denote the fitted values of the allometry parameters. Packard and Boardman [238] investigate the regression of data to a three-parameter power law that does not pass through the origin

$$Y = Y_0 + \overline{a}X^{\overline{b}} + \eta. \tag{3.2}$$

In the near isometry case where $\overline{b} \approx 1$ linear regression analysis is approximately valid and additive fluctuations provide a satisfactory representation of the statistical variability. On the other hand, when the allometry exponent is substantially different from one, the determination of the nature of the fluctuations requires preliminary statistical analysis, see, for example, Packard [237]. This is one reason that the data has been logarithmically transformed historically, before statistical analysis is carried out.

Packard and Boardman [238] emphasized that the additive form of fluctuations in ARs is quite different from the situation involving the logarithmically transformed data. For the transformed data, introducing additive random fluctuations yields

$$\log Y = \log \overline{a} + \overline{b} \log X + \eta, \tag{3.3}$$

and the empirical constants $\overline{a}$ and $\overline{b}$ are then fit to the transformed data. In this case, in terms of the original AR, we obtain

$$Y = \overline{a}e^{\eta}X^{\overline{b}}, \tag{3.4}$$

in which case the fluctuations in functionality are exponentially amplified, through $e^{\eta}$. It is evident that when the fluctuations are considered to be focused in the allometry coefficient, we can define the random allometry coefficient by

$$a = \overline{a}e^{\eta}, \tag{3.5}$$

in which the fluctuations are multiplicative. The multiplicative character of the fluctuations implies that the influence of the random variations on the allometry coefficient can be amplified far beyond their additive cousins.

Packard and Boardman go on to emphasize that the focus of the research on logarithmically transformed data is to characterize patterns of variation in morphology, physiology and ecology in organisms. This research spans a broad range in body size, in an attempt to identify underlying principles in the design of biological networks, see, for example, Brown et al. [50] and references therein. They go on to assert that many of the patterns identified by this research are inaccurate and misleading and these mischaracterizations likely contribute to the ongoing debate about ways in which animals evolve.

We saw that additive fluctuations in the logarithmically transformed data are equivalent to multiplicative fluctuations in the original data. So the important question is whether or not it is necessary to perform the logarithmic transformation at all. Packard and Boardman [238] point out that the original motivation for carrying out the logarithmic transformation was to linearize the equations thought to represent the data and therefore facilitate the implementation of graphical and statistical analysis [247, 300]. In particular, the transformation allows the use of linear regression to fit the allometry parameters. However, they go on to show the biasing problems associated with such transforms, using computer generated data sets and caution that with the present day computer software for fitting nonlinear equations, linearization is no longer a sufficient rationale for logarithmic transforming data.

So the question arises whether or not there are other reasons to transform the data?

Kerkhoff and Enquist [165] strongly disagree with the conclusions reached by Packard [237] that standard methods for fitting allometry models produce "biased and misleading" results. They point out that most biological phenomena are inherently multiplicative [110, 114] and it is the proportional, rather than absolute, variation that matters. Moreover the multiplicative influence of fluctuations seen in the logarithmically transformed data is often misinterpreted as bias [300, 386]. They [165] maintain that the multiplicative error model is an appropriate feature, rather than a defect, of standard allometry analysis. Recent research suggests that geometric error, resulting from multiplicative fluctuations, should be the default standard for parameter estimation in biology [114, 130] and not the more traditional additive error. We explore elements of the dramatic differences between additive and multiplicative fluctuations subsequently in the present chapter and in Chapter 4.

## 3.2 Phenomenological distributions

We now consider the statistics of the fluctuations in an AR, using published data. The observational data relating the average energy in watts, expended by an animal from a given species, the basal metabolic rate, to the average total body mass of that species in kilograms, for 391 species of mammal, is plotted in Figure 3.1, see also Heusner [145], as well as, Dodds et al. [89]. A fit of the interspecies AR given by equation (1.26) to these data that minimizes the mean-square error is a straight line on double logarithmic graph paper and was found to have slope $\overline{b} = 0.71 \pm 0.008$ so that empirically the allometry exponent falls in the interval $2/3 < \overline{b} < 3/4$, with the allometry coefficient $\overline{a} = 0.02$. As West and West [355] noted, Heusner [144] had questioned Kleiber's fitted value of 3/4 and concluded from his own data analysis that this value of 3/4 was a statistical artifact. Feldman and McMahon [100] agree with Heusner's conclusions, but suggest that there was no compelling reason for the intraspecies and interspecies allometric exponents to be the same. After all, the intraspecies exponent is predicted to be 2/3 by geometric similarity and the interspecies exponent is predicted to be 3/4 by McMahon's elastic similarity model, as subsequently discussed in Chapter 4.

### 3.2.1 Allometry coefficient fluctuations

The metabolic allometry parameters are fixed by the fit shown in Figure 3.1 to the values $\overline{a}$ and $\overline{b}$. However, we [353] were struck by the significant variability around the line segment provided by the interspecies AR model with the fitted parameters. Therefore, we speculated that it is not unreasonable to reinterpret these fluctuations as random variations in either the allometry coefficient, the allometry exponent, or both. Note that since the AR is empirical we are free to choose the allometry parameters in any way that furthers our understanding of the process. In the metabolic case we identify the parameters as random variables. If we arbitrarily assume the fluctuations to be in the allometry coefficient as in equation (3.4) we can define the scaled allometry coefficient:

$$\frac{a}{\overline{a}} = e^{\eta} = \frac{\langle B_i \rangle}{\overline{a} \langle M_i \rangle^{\overline{b}}}, \tag{3.6}$$

so that each data point in the $(\langle B_i \rangle, \langle M_i \rangle)$-plane yields a single value of the allometry coefficient. Note that equation (3.6) has been interpreted as the residual variation in

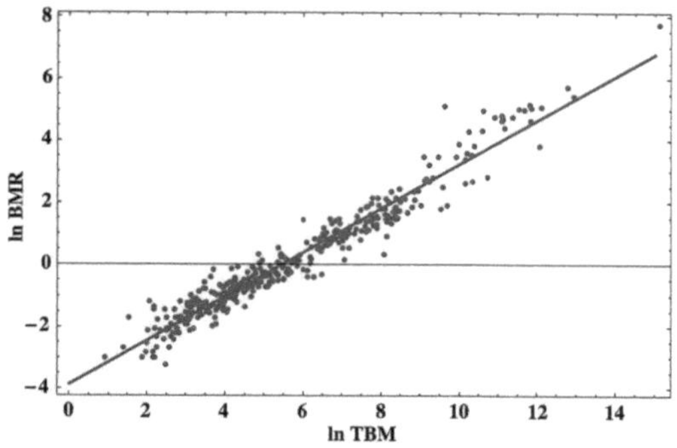


**Figure 3.1:** The linear regression to equation (1.26) for Heusner's data [145] is indicated by the line segment. The slope of the line segment is $0.71 \pm 0.008$. (Reproduced from [350] with permission).

$\langle B_i \rangle$ by Dodds, Rothman and Weitz [89], and such an interpretation assumes the linear regression models a literal response of $\langle B_i \rangle$ to fluctuations in $\langle M_i \rangle$. However, in a complex network, such a linear response does not necessarily occur, since there can be independent fluctuations in both $\langle B_i \rangle$ and $\langle M_i \rangle$ resulting in what Warton et al. [327] call equation error; also known as natural variability, natural variation and intrinsic scatter in which the AR is not predictive, but instead summarizes vast amounts of data [374]. This natural variability is manifest in fluctuations in the space of the allometry parameters $(a, b)$.

Heusner [145] considered the allometry coefficient to be the "remaining mystery" for the BMR AR, but few considered his arguments seriously; most investigators focused their attention on the reductionist modeling of the allometry exponent [365]. The normalized fluctuations in the allometry coefficient, with the empirical value of the allometry exponent fixed at $\overline{b} = 0.71$ are clearly evident in the data depicted in Figure 3.2, determined by equation (3.6).

West and West [350] calculate the statistical distribution for the random allometry coefficient determined from equation (3.6), under the assumption that $\overline{b}$ and $\overline{a}$ are fixed. Dodds, Rothman and Weitz [89] considered these same fluctuations, but with $\overline{b}$ replaced with 2/3 and interpreted them as the residual variations in the metabolic rate. The latter authors concluded from their analysis that the fluctuations in the residuals have a log-normal distribution. However, we [350] find a different distribution, using equation (3.6), that we now discuss.

The statistical distribution for the random allometry coefficient, using equation (3.6) under the assumption that $\overline{b}$ is fixed, is determined by West and West [350] to be IPL. The variability in the allometry coefficient determined by the data is partitioned into twenty equal sized bins in the logarithm of the allometry coefficient. A histogram is then constructed by counting the number of data points within each of the bins, as indicated by the dots in Figure 3.3. The solid line segment in this figure is the best fit to these twenty numbers, with minimum mean-square error. The

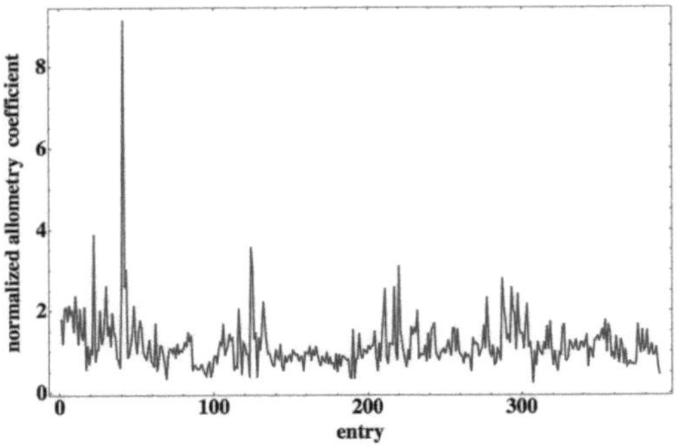


**Figure 3.2:** The fluctuations around the AR using Heusner's data set [145] are shown and the line segments connect the data points to aid the eye in estimating the variability. Note that this figure depicts the erratic value of the allometric coefficient $a$ in equation (3.5), with a single value for each $(B, M)$-pair for the normalized data with $\overline{a} = 0.02$ and a fixed $\overline{b} = 0.71$. (From [350] with permission.)

functional form for the histogram is indicated by the curve in Figure 3.3 [349, 350] and the quality of the fit to the diversity data is determined by the correlation coefficient $r^2 = 0.98$. The normalized histogram $G(\ln a')$ on the interval $(0, \infty)$ using the transformation

$$G(\ln a')d\ln a' = P(a')da' \tag{3.7}$$

gives the empirical PDF:

$$P(a') = \frac{\alpha}{2}\begin{cases} \frac{1}{a'^{1-\alpha}}; & a' \leq 1, \\ \frac{1}{a'^{1+\alpha}}; & a' \geq 1 \end{cases} \tag{3.8}$$

and the best fit value of the exponent is $\alpha = 3.28$, yielding a standard deviation 0.017 in essential agreement with the empirical data. Equation (3.8) becomes an IPL PDF asymptotically. Note that this coefficient differs from the best fit value given in the caption of Figure 3.3, but this results in only the negligible change in the value of the quality of fit parameter $r^2$ of 1%.

The same IPL form is obtained with $\alpha = 3.89$, with a quality of fit $r = 0.96$, using the avian BMR data of McNab [212, 213] for 533 species of bird. The distribution of the deviations from the AR, for both the avian and mammalian data sets, fall off as IPLs on either side of $a = \overline{a}$. Equation (3.8) quantifies the qualitative argument used earlier to associate IPL PDF's with multiplicative fluctuations. These heavy tailed PDFs strongly suggest a natural inclination towards variability or diversification in the macro-evolution of species; a diversity that provides a fractal advantage, as discussed in Section 1.4.3.

The asymptotic form of equation (3.8) provides a Pareto PDF and was discussed by Mandelbrot [201] in terms of self-similarity of the statistics of data. The difference between Gaussian statistics and those of Pareto is remarkable. The former is completely

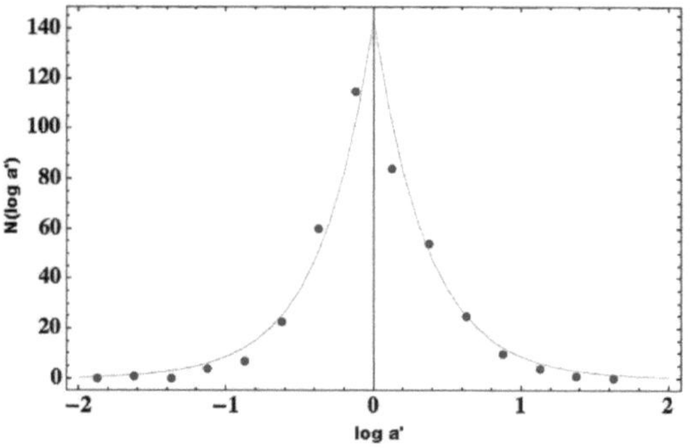


**Figure 3.3:** The histogram of the deviations from the prediction of the AR using the data depicted in Figure 3.2 partitioned into 20 equal sized bins in the logarithm of the normalized variable $a' = a/\overline{a}$. Here $\overline{a} = 0.02$ and $\overline{b} = 0.71$. The solid line segment is the best fit of equation (3.8) to the twenty histogram numbers, which yields the power-law index $\alpha = 2.79$ and the quality of the fit is measured by the correlation coefficient $r^2 = 0.97$. (From [350] with permission.)

characterized by the mean and variance, in which case, like the LFE, would suggest nature has a preference for a single value of the allometry coefficient. On the other hand, the extended tail of the Pareto PDF, indicates that the variance can diverge for $\alpha < 2$ and even the mean can diverge for $\alpha < 1$.

The Pareto PDF characterizes data having a great deal of variability and is dominated by fluctuations that would be identified as outliers, if viewed from the Gauss perspective. Consequently, the probability that the allometry coefficient exceeds the value $A'$ is given by using equation (3.8) to be

$$\Phi(A') = \int_{A'}^{\infty} P(a')da' \propto \frac{1}{A'^{\alpha}} \tag{3.9}$$

for $A' \gg 1$. The interspecies AR describes a trait (functionality) across multiple species and the Pareto PDF equation (3.9) characterizes the variability (diversity) of that trait across species. Therefore, the observed variability in the AR is a consequence of the intermittent statistical fluctuations in the allometry coefficient, according to this approach.

### 3.2.2 Allometry exponent fluctuations

A complementary phenomenological approach, which seems equally reasonable mathematically, is to assume that the allometry coefficient is a constant $\overline{a}$ and the variation in the AR is due to the random nature of the allometry exponent $b$. We write the fluctuations in the allometry exponent, for the data set in Figure 3.1, as

$$\eta_i = \frac{\ln(\langle B_i \rangle / \overline{a})}{\ln \langle M_i \rangle} - \overline{b}. \tag{3.10}$$

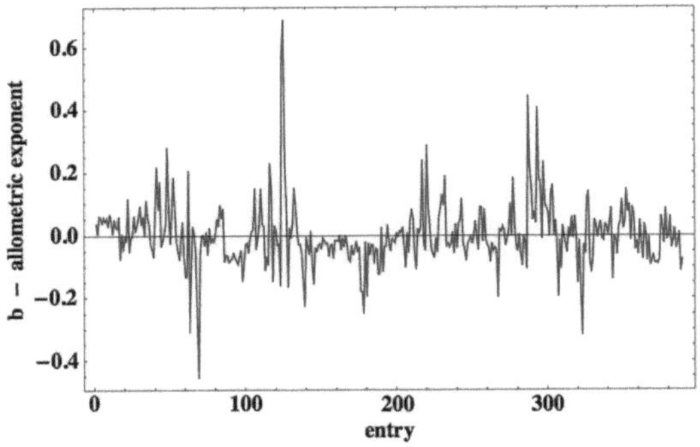


**Figure 3.4:** The line segments connect Heusner's data [145] points to aid the eye in estimating the variability. Note that this figure depicts the erratic value of the zero-centered allometry exponent $\eta_i$ in equation (3.10). (From [350] with permission.)

If we again assume $\overline{b} = 0.71$ and $a = \overline{a} = 0.02$, then equation (3.10) provides us with the statistical fluctuations in the allometry exponent shown in Figure 3.4.

Equation (3.10) provides us with the statistical fluctuations in the allometry exponent depicted in Figure 3.4, which are used to construct a histogram exactly as done previously for the allometry coefficient, starting from the same raw data. The solid line segment in Figure 3.5 is the best fit to the twenty numbers of the histogram, with minimal mean-square error. The functional form for the histogram of deviations from the allometry exponent $\overline{b}$ is determined by the curve in Figure 3.5 and the quality of the fit to the histogram is determined by the correlation coefficient $r^2 = 0.97$. The histogram data in Figure 3.5 are fit by the Laplace PDF

$$\Psi(b) = \frac{\beta}{2} \exp[-\beta|b - \overline{b}|], \tag{3.11}$$

with the empirical value $\beta = 12.85$.

Note the dramatic differences in the two distributions for the allometry parameters given in Figures 3.3 and 3.5. The IPL PDF for the allometry coefficient can have diverging central moments for $\alpha < 2$ and emphasizes the variability of influence of the fluctuations, when they are all interpreted as being multiplicative in the coefficient. On the other hand, the Laplace PDF for the allometry exponent has all central moments finite and emphasizes the average as the dominant influence of the fluctuations, when they are all interpreted as being additive in the exponent.

Of course, there is a third option, and that is to fit a two-dimensional PDF to the scattering of the parameter values in the $(a, b)$ plane. We have not followed that steeper road here, but suggest that it might be worth exploring for a young researcher seeking a potentially useful project. Note that the variables in such a two-dimensional PDF need not be independent of one another, as is so often assumed. In fact the vast majority of researchers treat the allometry coefficient as if it were an unimportant normalization constant, devoid of useful information. We now put that assumption to the test

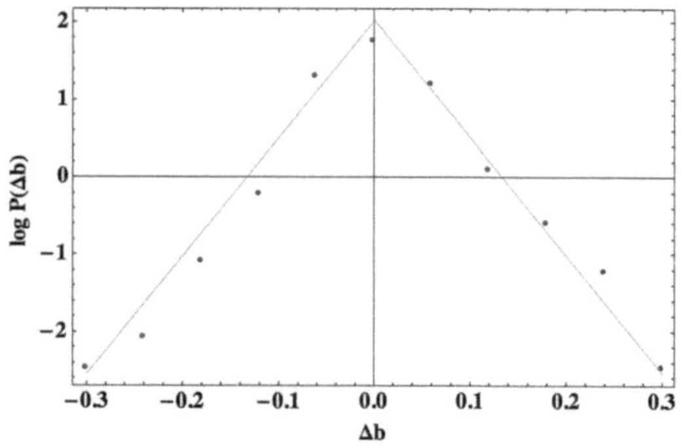


**Figure 3.5:** The histogram of the deviations from the prediction of the AR, using the allometry exponent fluctuation data partitioned into 20 equal sized bins. The solid line segment is the best fit of equation (3.11) with $\Delta b \equiv b - \overline{b}$, to the twenty histogram numbers, and the quality of the fit is measured by the correlation coefficient $r^2 = 0.97$. (From [350] with permission.)

and turn our attention to determining the possible interdependence of the allometry coefficient and exponent.

### 3.2.3 Other scaling statistics

The tent shaped distribution of Laplace in Figure 3.5 also arises using a different approach to quantifying the variability of BMR. Labra et al. [176] investigate BMR fluctuations by considering $O_2$ volume time series and examining the scaling of the high frequency fluctuations across species. They determined empirically that the standard deviation in the BMR is proportional to the average BMR by the scaling law:

$$\mathrm{SD}_{\mathrm{BMR}} \propto \langle B \rangle^{\lambda}, \tag{3.12}$$

and the empirical exponent is determined to be $\lambda = -0.352 \pm 0.072$. Thus, using the variance of the BMR as a measure of complexity, we see that the complexity decreases as an inverse power of the average BMR. Note that this is an application of an empirical law developed by Taylor [313] to model the diversity of new species and is discussed in Section 3.2.4.

On the other hand, Labra et al. [176] determined that the standard deviation of the BMR data is proportional to a power of the average TBM:

$$\mathrm{SD}_{\mathrm{BMR}} \propto \langle M \rangle^{\gamma} \tag{3.13}$$

and the empirical exponent, in this case, is determined to be $\gamma = -0.241 \pm 0.103$. Here again the complexity is seen to decrease as an inverse power of the TBM. Taken together, it is evident that the complexity, as measured by the variance in the BMR, is more sensitive to increasing average BMR, than it is to increasing average TBM.

Consequently, combining the two empirical expressions for the standard deviation of the BMR data, equations (3.12) and (3.13), yields

$$\langle B\rangle \propto \langle M\rangle^{\gamma/\lambda}, \tag{3.14}$$

and from the empirically determined values of the independent parameters, their ratio is $\gamma/\lambda = 0.69$. The value of the exponent in equation (3.14) is not very different from the many empirical fits made to the AR data directly. They [176] determine that all the species they studied show the same invariant distribution of average BMR fluctuations, regardless of the difference in their phylogeny, physiology and body size. The distribution has the form equation (3.11) with the independent variable given by the fluctuations in the average BMR denoted by $\Delta B$ and $\beta = \sqrt{2}/\mathrm{SD}_{\mathrm{BMR}}$; in terms of the scaling variable $B_{sc} \equiv \sqrt{2}\Delta B/\mathrm{SD}_{\mathrm{BMR}}$ curves for 12 different species collapse onto a single universal curve.

It is worth emphasizing that one of the simplest measures of complexity of a statistical process is the standard deviation. Consequently, equations (3.12) and (3.13) provide independent measures of the complexity of physiological system using BMR (functionality) and TBM (size), respectively. These two measures of complexity are then combined to obtain the empirical AR given by equation (3.14), where the complexity is no longer evident. This is another piece of evidence supporting the thesis that allometry is a manifestation of a system's complexity, which we take up further in the next subsection.

### 3.2.4 Taylor's Law

Taylor's Law is also known as the power curve in the ecology literature [139] and was discovered by Taylor [313] in his determination of the number of new species of beetle that can be found in a given plot of ground and thereby provide a measure of species diversity [346]. He sectioned off a plot of land into a checker board of parcels and sampled each one in the same way to determine the parcel-specific number of new species of beetle. He then aggregated this number across parcels to obtain an average number of new species, denoted by $\langle N\rangle$, with the variance around that number denoted by $\langle N^2\rangle - \langle N\rangle^2$. After this first collection he then partitioned his land into smaller parcels and repeated the procedure to obtain new values of the mean and variance. He repeated this procedure of making smaller patches a number of times and assumed a scaling form for the relation between the mean and variance

$$\langle N^2\rangle - \langle N\rangle^2 = a\langle N\rangle^b, \tag{3.15}$$

which he fit to the data obtained from multiple applications of his data sampling procedure. These data provide a direct measure of the complexity of the underlying process in terms of the variance.

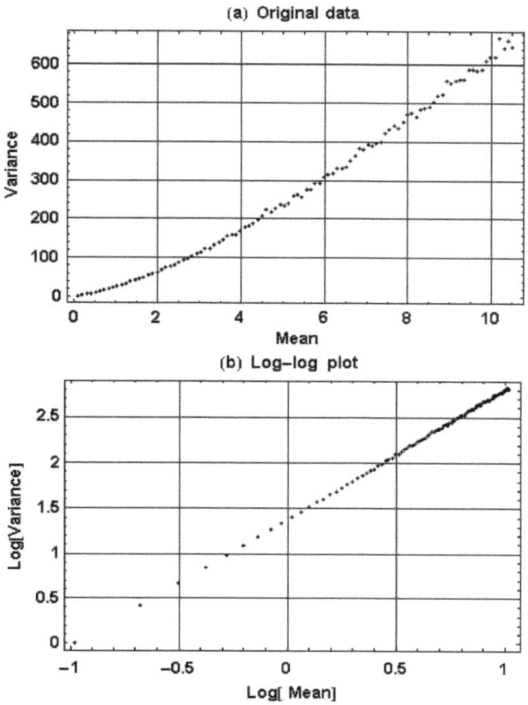


**Figure 3.6:** (a) Computer-generated data that satisfies Taylor's power-law relation between the standard deviation and mean is plotted. (b) The same data used in (a), after taking the logarithm of the variance and the logarithm of the mean, is plotted. It is evident that the log–log graph yields a straight line when the variance and average satisfy the Taylor's Law equation (3.15).

A hypothetical data set satisfying equation (3.15) is plotted in Figure 3.6 (a). The variance on the vertical axis grows nonlinearly with the increasing average on the horizontal axis, assuming fluctuations such as might be observed in real data sets. If the data in Figure 3.6 (a), in fact, satisfy the power-law relation of Taylor, then we can take the logarithm of the data to obtain the graph shown in Figure 3.6 (b). The dominant behavior is a direct proportionality between the two logarithmically transformed variables. The proportionality constant is the slope of the curve, which algebraically is identified with the power-law index in equation (3.15).

As explained in [346], Taylor was able to exploit curves such as in Figure 3.6 (b) using the two parameters. If the slope of the curve and the intercept are both equal to one, $a = b = 1$, then the variance and the mean are equal. This equality is only true for a Poisson distribution, which, when it occurs, allows one to interpret the number of new events (species) as being randomly distributed, which in his case would be over the field with the number of species in any one parcel being completely independent of the number of species in any other parcel. If, however, the slope of the curve was less than unity, $b < 1$, the number of new events (species) appearing in the data sets (parcels) can be interpreted to be quite regular. The spatial regularity of the number of species was compared with the trees in an orchard and given the name *evenness*. Finally, if the slope is greater than one, $b > 1$, the number of new events (species) is clustered in space, like disjoint herds of sheep grazing in a meadow.

Of particular interest to us here was the mechanism that Taylor and Taylor [314] subsequently postulated to account for the heuristic AR:

> We would argue that all spatial dispositions can legitimately be regarded as resulting from the balance between two fundamental antithetical sets of behavior always present between individuals. These are, repulsion behavior, which results from the selection pressure for individuals to maximize their resources and hence to separate, and attraction behavior, which results from the selection pressure to make the maximum use of available resources and hence to congregate wherever these resources are currently most abundant.

Consequently, they postulated, using the language of mechanical forces, that it is the balance between the attraction and repulsion, migration and congregation, that produces the interdependence of the spatial variance and the average population density. This could be interpreted, more accurately, as being an imbalance in complexity or information, such that there is an information force, resulting from an imbalance in complexity, here interpreted as selection pressures, acting in opposition to one another. The more complex, that is, the increasing number of newer species of beetle, moves into the physical region of the less complex, or older, species of beetle and eventually replaces them.

The slope of the power curve being greater than one is consistent with an asymptotic IPL PDF, which entails the clustering or clumping of events. This clustering is due to the fractal nature of the underlying dynamics. Willis [379], some 40 years before Taylor articulated his law, established the IPL form of the number of species belonging to a given genera. Willis used an argument associating the number of species with the size of the area they inhabit. It was not until the decade of the 1990s that it became clear to more than a handful of experts that the relationship between an underlying fractal process and its space filling character obeys a scaling law. It is this scaling law that is reflected in the AR, between the variance and the average in Taylor's Law. This law was extended to the determination of scaling in time series by Taylor and Woiwold [315] and applied in a physiology context by West [346].

An exhaustive, user-friendly, primer on the statistical methods used in the analysis of Taylor's Law is given in a paleoecological context in *Surveying Natural Population* by Hayek and Buzas [139]. What is remarkable to me is that, although Taylor's original paper has over one thousand citations and his law has been confirmed for hundreds of species in field observations and laboratory experiments, his law is discussed by Hayek and Buzas under the name *The Power Curve*, without acknowledging Taylor's contribution. More recently the origin of the statistics manifest in the relation between the variance and mean has been explored, using the methods of equilibrium statistical physics. In their analysis, Fronczak and Fronczak [103] consider the $N$ in equation (3.15) to be given by the number of: corn borers living on a given plant; turnovers of a given stock; birds of a given species observed in a particular area at a given time; cars passing a recording device at a given time, traveling in a given direction; and gene structures spanning a whole chromosome. Their intent in providing such a variety of data sets for analysis was to emphasize the generality of their approach as being commensurate with the universality of the law. Their arguments were based on statistical physics, focusing on the complexity of phenomena, as measured by the PDF and did not address any specific system.

Taylor's Law has recently been applied to tornado outbreaks in the United States [318]. In this analysis Tippett and Cohen [318] determined that the variability in tornado outbreaks for the time period 1954 to 2014 satisfied equation (3.15), with the linear regression fit yielding parameter values $b = 4.3 \pm 0.44$ and $\ln a = -6.74 \pm 1.12$. The value of exponent $b$ is noteworthy because in most ecological applications it rarely exceeds 2. Given that the variance is a measure of the complexity of the underlying process, this value would strongly suggest that the macrostructure of a tornado is significantly more complex than most ecological systems.

### 3.2.5 Paleobiology and scaling

Phenomena possessing properties with intermittent fluctuations, described by IPL statistics, are apparently ubiquitous [342] and those within paleobiology are no exception. Bak and Boettcher [14] interpreted Charles Lyell's [193] *uniformitarianism* as meaning that all geologic activity should be explainable in terms of readily available processes working at all times and all places with the same intensity. They go on to assert that the existence of earthquakes, volcanic eruptions, floods and tsunamis all indicate that the physical world is far from equilibrium. Moreover the intermittent nature of the paleontological record indicates that macro-evolution is also out of equilibrium and consequently the IPL statistics are possibly suitable for their description.

Eldredge and Gould [93] argued that punctuated change dominates the history of life and that relatively rapid episodes of speciation constitute biological macroevolution. The intermittency of speciation in time has been explained by one group as *punctuated equilibria* [92] and has been indirectly related to fractal statistics by identifying it as a self-organized critical phenomenon [304]. In the self-organized criticality model of speciation, Bak and Boettcher [14] associate an avalanche of activity with exceeding a threshold and the distribution of returns to the threshold with a "devil's staircase", having a distribution of steps of stasis lengths given by the IPL PDF, $T^{-\gamma}$ with $\gamma = 1.75$. Moreover, as explained by Sneppen et al. [304], the number of genera $N$, with a lifetime $T$ can be fitted very well to an IPL $T^{-\beta}$ with $\beta \approx 2$ [263]. More recently, Rikvold and Zia [271] put forward an explanation of punctuated equilibrium, based on

$1/f$ noise in macro-evolutionary dynamics that also yields an IPL PDF for the life time of ecological communities, having an IPL index of $\beta = 2$.

Solé et al. [308] analyzed the statistics of the extinction fossil record (time series) and determined that the power spectrum has the form

$$S(f) \propto 1/f^{\mu}, \tag{3.16}$$

with $0 < \mu < 2$. They find from data that $0.80 \leq \mu \leq 0.90$ and reason that these values support the self-organized criticality interpretation of extinction. On the other hand, Plotnick and Sepkowski [250] also find a $1/f$ power spectrum, with an IPL index for extinction consistent with Solé et al. [308] and indices for species generation of approximately half that for extinction. However, the latter authors conclude that their results are incompatible with self-organized criticality and instead are compatible with multifractal self-similarity in both the extinction and generation records.

### 3.2.6 Urban variability

Bettencourt et al. [36] in their study of urban scaling constructed the metric

$$\eta_i = \ln\left[\frac{Y_i}{aX_i^b}\right], \tag{3.17}$$

which they called the Scale-Adjusted Metropolitan Indicators (SAMIs). They used for $Y_i$ the observed value of the measure of innovation, wealth or crime for each city $i$ with population $X_i$. They find that a Laplace distribution provides an excellent fit to the normalized SAMI histogram for the statistical residuals $\eta_i$ across different cities. However, they make the bold assumption that the allometry exponent is approximately universal.

Quite independently and contemporaneously an analogous measure was devised by us [352] for the relative variation in the allometry parameters for the metabolic AR, see Chapter 2. We [352] reasoned that since there are independent fluctuations in $X$ and $Y$, the AR resulted in what Warton et al. [327] call equation error; also known as natural variability, natural variation and intrinsic scatter. Considering that ARs are not predictive, but instead summarize vast amounts of data [374], this natural variability was interpreted as fluctuations in the modeling allometry parameters $(a, b)$. Denoting the fitted values of the random parameters as $\overline{a}$ and $\overline{b}$, if the fluctuations are assumed to be contained in the allometry coefficient we [352] define the residual in the allometry coefficient:

$$\eta_i = \ln\left[\frac{\langle Y_i\rangle}{\overline{a}\langle X_i\rangle^{\overline{b}}}\right]. \tag{3.18}$$

The numerator and denominator in equation (3.18) are measured independently and in the case we were investigating $\langle Y_i\rangle$ is the average BMR and $\langle X_i\rangle$ is the average TBM.

The statistics of the normalized allometry coefficient given by equation (3.6), were determined by least squares fitting to the data to be given by a two-sided IPL PDF depicted in Figure 3.3.

On the other hand, when the fluctuations are assumed to be constrained in the allometry exponent, we define the statistical residual as

$$\eta_i = \frac{\ln(\langle Y_i \rangle / \overline{a})}{\ln \langle X_i \rangle} - \overline{b}. \tag{3.19}$$

In our analysis, the allometry coefficient and exponent were held fixed, so that the parametric fluctuations fitted by a histogram gave the best fit to be that of a Laplace PDF centered on the fitted value of the allometry exponent as depicted in Figure 3.5. This latter result is consistent with the PDF obtained using the SARI method of Bettencourt et al. [36].

Both research groups reach the conclusion that the Laplace PDFs for the statistics of the allometry exponent imply the IPL PDF in the size of the network. This overlap of interpretation was reached in spite of the fact that in one case the data consisted of independent measures of BMR and TBM, which is a convex AR ($b < 1$), and in the other case the data consisted of independent measurements of city economic quantities and populations in a given year, which is a concave AR ($b > 1$). The convergence of conclusions reached in these two studies suggests the necessity for statistical measures being considered as foundational for understanding allometry in general, as had been argued previously [355].

## 3.3 Are ARs universal?

A number of theoretical studies [17, 95, 366] maintain that living networks ought to have universal scaling laws. In the AR context a universal scaling law is one in which the allometry exponent has a constant value and the allometry coefficient is an unimportant normalization constant. In a remarkable review, West and Brown [371] pose the fundamental question:

> Do biological phenomena obey underlying universal laws of life that can be mathematized so that biology can be formulated as a predictive, quantitative science?

They qualify their affirmation of a quantifiable universal law in terms of a coarse-graining behavior of living systems and offer scaling as an exemplar of such universal behavior in the physical sciences. Their review focuses on the scaling behavior of ARs in living systems and stresses the significance of the metabolic allometry exponent having the value 3/4. They maintain that the 'quarter-power' scaling laws in biology reflect fundamental underlying constraints and explain those constraints in terms of fractal networks and energy minimization. We subsequently give only a brief review of their model [365] of nutrient transport, because of the large number of previous reviews and critiques of their theory, see for example, [9, 78, 83, 96, 174, 175, 298, 283, 355], along with the many commentaries on the data not yielding the value $b = 3/4$, see, for example, [39, 89, 119, 373].

In this section we focus on their specific requirement that $b = 3/4$ in order for there to be a universal law for metabolic ARs. West and Brown [371] disagree with the criticisms of the 3/4-scaling exponent and maintain that this value is overwhelmingly supported by the data; while at the same time acknowledging the trend toward $b = 2/3$ in mammals of 10 kg or less. A meta-analysis of 127 published metabolic ARs for birds, mammals, fish, reptiles, amphibians and anthropoids made by White et al. [374] supports neither $b = 2/3$ nor $b = 3/4$. We follow [356] and review evidence concerning the constancy of the allometry exponent, using data and theory relevant to the covariation of the allometry coefficient (metabolic level) and the allometry exponent. The existence of this covariation is not consistent with the universal metabolic AR sought above.

### 3.3.1 Covariation of allometry parameters

Glazier [119] points out that the resource-transport models, as they are presently formulated, cannot explain the variability in the metabolic allometry exponent $b$ observed in organisms. He observes that these differences in $b$ are seen within groups of related species and even within a given species, where one would not expect them, based on the minimal differences in resource-distribution systems. The scaling of metabolic rates has long been a controversial topic, specifically with regard to whether the allometry exponent has a particular value. The empirical metabolic AR within a given taxonomic group relates the average BMR $\langle B_i \rangle$ to the average TBM $\langle M_i \rangle$ for species $i$:

$$\langle B_i \rangle = a \langle M_i \rangle^b . \tag{3.20}$$

There is a great deal of variability in the values of the allometry parameters, obtained by fitting data to equation (3.20). But rather than joining the controversy over a single value of the allometry exponent, as had been done by almost every previous investigator, Glazier [117, 119] decided to pursue an entirely different approach.

He [119] posited the metabolic-level boundaries (MLB) hypothesis as one way to incorporate ecologic lifestyle and activity level into the metabolic intensity of specific groups of organisms. He also critiqued other models that had been proposed, but as he makes clear, his is the only model capable of explaining the wide variation in $b$ that has been observed in both taxa and among physiologic states. As those other models are presently constituted they can either explain the variation of metabolic scaling, due to taxonomic variation, or to physiologic state, but not both.

In the MLB theory the emphasis shifts from average tendencies to boundary conditions. One extreme boundary constraint is the surface-area limits on fluxes of metabolic resources, waste and heat; all of which scale as $\langle M\rangle^{2/3}$. Another extreme boundary constraint is the volume limits on power production, which scales as $\langle M\rangle$. Both these constraints must be satisfied by $b$. Next, it is not sufficient to consider the variation with size in the exponent alone, but following the suggestion of Heusner [145] the importance of the allometry coefficient must be taken into account. As we have repeatedly observed, this parameter is usually ignored as irrelevant. Calder [57] contends that in the absence of a *theory of allometry* the AR is merely a statement of correlation between two quantities. This perspective is consistent with the interpretation introduced by Huxley [152], along with the first mathematical theory of ARs that the allometry coefficient "has no biological or general significance".

Aside from isolated observations made by Gould [128] and Heusner [144], it is only recently that the elevation of the fitted data in the $(B,M)$-plane, the metabolic level ($\log a$), has been considered important, due to its systematic variation with slope [117, 119]. The third and final shift in focus Glazier made concerns the ecological effects that influence both allometry parameters and not just the slope of the fitting curve in the $(B,M)$-plane. The MLB hypothesis systematically incorporates the above three shifts in focus, by reasoning that $b$ should be negatively correlated to the metabolic level in resting organisms and positively correlated in actively moving animals. He argues that $b = 2/3$ for low metabolic levels and $b = 1$ for high metabolic levels, resulting in a V-shaped dependency of the allometry exponent on the allometry coefficient.

The empirical testing of the MLB hypothesis was based on the most recent and comprehensive data sets on metabolic scaling that were available early in this decade. The dots in Figure 3.7 are the metabolic rates for mammals supplied to us by Glazier in a private communication and are taken from Figure 6D of [119]. This figure is only one of a number of such graphs in his paper that clearly establish the V-shaped covariation of $b$ with $\log a$ resulting from Glazier's data processing and explained using the MLB hypothesis.

### 3.3.2 The principle of empirical consistency

As previously mentioned, one cannot uniquely attribute random fluctuations to the average BMR, or to the average TBM; a proper theory must achieve consistency of results independently of which variable is assumed to be the source of the fluctuations [349], or the degree to which they each contribute to the variability. Alternatively, the random fluctuations may be assumed to reside in the allometry parameters in the $(a,b)$-plane instead of in the average physiologic variables in the $(B,M)$-plane. This arbitrariness is a consequence of the allometry model being statistical and is not reductionistically determined.

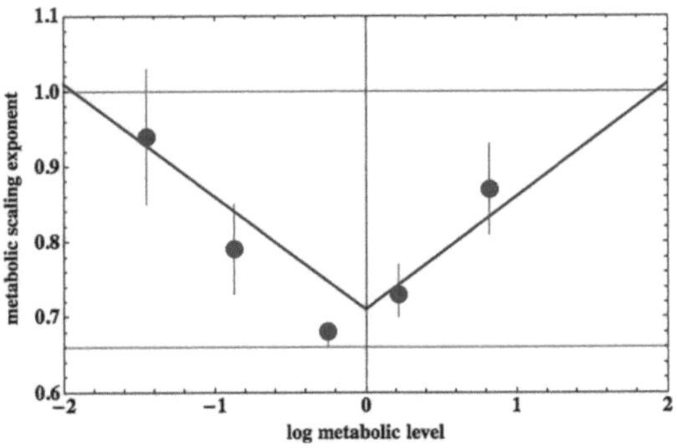


**Figure 3.7:** The metabolic scaling exponent $b$ is graphed versus the logarithm of the metabolic level $a$, among mammals at different activity levels. The dots are data points supplied by Glazier in a private communication and depicted in Figure 6D of Glazier [119] with the ±95% confidence intervals indicated. See Glazier [119] for the details of the data analysis. The theoretical relation between the allometry exponent $b$ and the allometry coefficient (logarithm of the metabolic level $\log_{10} a'$ given by equation (3.30)) is given by the heavy line segments. The upper line segment denotes $b = 1$ and the lower one $b = 2/3$. (From [349] with permission.)

The PDF for the allometry exponent $b$, with the allometry coefficient held fixed at $a = \overline{a}$, was determined to be that of Laplace [349]:

$$\Psi(b;\overline{a}) = \frac{\beta}{2} e^{-\beta|b-\overline{b}|}. \tag{3.21}$$

The empirically fit parameters are $\beta = 12.85$, $\overline{b} = 0.71$, with the quality of fit $r^2 = 0.97$. Using the same data, it is also possible to determine the PDF for the allometry coefficient $a$, with the allometry exponent held fixed at $b = \overline{b}$, to obtain an IPL PDF. The PDF, in terms of the normalized variable $a' = a/\overline{a}$ [349], is:

$$P(a';\overline{b}) = \frac{\alpha}{2} \begin{cases} \frac{1}{a'^{1-\alpha}} & \text{for } a' \leq 1, \\ \frac{1}{a'^{1+\alpha}} & \text{for } a' \geq 1, \end{cases} \tag{3.22}$$

the same as equation (3.8). The empirically fit parameter is $\alpha = 2.79$, with the quality of fit $r^2 = 0.98$. It was mentioned that the same distributions, with slightly different parameter values, are obtained using the avian data of McNab [213].

A given fluctuation in the $(a, b)$-plane is equally likely to be the result of a random variation in the allometry coefficient or in the allometry exponent and therefore the probability of either occurring should be the same:

$$\Psi(b;\overline{a})db = P(a';\overline{b})da', \tag{3.23}$$

which we [356] referred to as the *Principle of Empirical Consistency* (PEC). If equation (3.23) is to be mathematically valid the allometry parameters must be functionally related, so we impose the constraint:

$$b = \overline{b} + f(a'). \tag{3.24}$$

The unknown function $f(a')$ is determined by inserting equation (3.24) into equation (3.23) to obtain the differential equation for the allometry parameters in terms of the ratio of the empirical PDFs:

$$\frac{db}{da'} = \frac{df(a')}{da'} = \frac{P(a';\overline{b})}{\Psi(\overline{b} + f(a');\overline{a})}. \tag{3.25}$$

Equation (3.25) defines a relation between the allometry parameters through the function $f(a')$ in terms of the empirical PDF's. Inserting the empirical PDFs into equation (3.25), yields the differential equation:

$$\frac{df(a')}{da'} = \begin{cases} \frac{\alpha}{\beta a'^{1-\alpha}} \exp[\beta|f(a')|] & \text{for } 0 < a' \le 1, \\ \frac{\alpha}{\beta a'^{1+\alpha}} \exp[\beta|f(a')|] & \text{for } a' \ge 1 \end{cases} \tag{3.26}$$

to be solved.

Integrating equation (3.26) by inspection, the values of $f(a')$, including a constant of integration $C$, in the indicated domains are

$$f(a') = C \begin{cases} -\ln a' & \text{for } a' \le 1, \\ \ln a' & \text{for } a' \ge 1. \end{cases} \tag{3.27}$$

Tailoring the solution to the metabolic level boundaries discussed by Glazier [119], we introduce constraints on the solution, such that the maximum value of the allometry exponent is $b = 1$ and the unknown function has the value at the boundaries

$$f(a') = 0.29 \quad \text{at } \log a' = \pm 2 \tag{3.28}$$

resulting in $C = 0.35$. Consequently equation (3.27) can be written in compact form:

$$f(a') = 0.15|\log a'|. \tag{3.29}$$

Thus, substituting equation (3.29) into equation (3.24) and noting that empirically $\overline{b} = 0.71$, the allometry exponent is given by the relation to the allometry coefficient

$$b = 0.71 + 0.15|\log a'| \tag{3.30}$$

which has the V-shaped form indicated with the solid line segment in Figure 3.7.

Equation (3.30) is consistent with the phenomenological expressions constructed from data by Glazier [119], using the MLB hypothesis for both intraspecies and interspecies ARs. The dots in Figure 3.7 are from the mammalian species data depicted in

Figure 6D of Glazier [119] and are well accounted for by equation (3.30). Note that the theoretical curve is not fit to these data points, but is given by equation (3.30) using empirical PDFs determined by a different data set. The above probability model capturing the statistical fluctuations in the allometry parameters, through the empirical PDF's, was made consistent with the MLB hypothesis [119] by having the solution to the differential equation satisfy the same boundary conditions.

The covariation of the allometry coefficient and allometry exponent, resulting from the MLB hypothesis, or the PEC, indicate that the metabolic ARs are not universal. If the metabolic ARs are not universal, and these were the ARs with the strongest evidence for a constant allometry exponent, there does not appear to be a compelling argument of universality of ARs in general. White et al. [374] put it eloquently in their conclusions:

> ...a century of science was distorted by trying to fit observations to an unsatisfactory surface law ($b = 2/3$). Given the apparent widespread acceptance and application of $b = 3/4$, it seems history is in danger of repeating.

## 3.4 Summary

In this chapter, we introduced a new way of modeling metabolic allometry, that being, to assume the form of the AR, but to introduce allometry parameters that can have statistical fluctuations. This approach results in an emphasis of the importance of the allometry coefficient, given its multiplicative placement in the AR and its IPL variability. Thus, the influence of the fluctuations in the allometry coefficient on functionality is enhanced for large animals and suppressed for small ones. At the same time the data reveal a relatively narrow Laplace distribution of values of the allometry exponent around its average value.

It should be emphasized that the covariation function given by equation (3.30) is arrived at from two very different perspectives. Glazier [119] developed a remarkable biological mechanistic model, which he used to fit model parameters to data. In shifting the focus from averages to boundary conditions he was able to determine the phenomenological V-shaped covariation function. We adopted a probabilistic argument requiring consistency between two empirical PDFs; one for each of the allometry parameters in the form of new principle, the PEC. The consistency requirement results in the theoretical covariation function given by equation (3.30).

Note that the two approaches do not contradict one another. One approach is similar in spirit to determining the normal distribution from the central limit theorem, whereas the other is analogous to a minimization argument constrained by an empirical mean and variance. The former establishes the general properties the data must have to obtain the normal distribution as a limit; the latter imposes consistency between the general properties of two PDFs.

It is also worth emphasizing that Taylor's Law is a direct measure of the complexity of a process, giving rise to the variance of an observable being proportional to the average value of the observable raised to a non-integer power in general. Taylor's Law entails an AR, as can be seen from the independent measurements of a process in terms of two variables $V_1$ and $V_2$, resulting in

$$\text{variance } V_1 \propto \langle V_1 \rangle^{h_1},$$
$$\text{variance } V_2 \propto \langle V_2 \rangle^{h_2}.$$

However, since the variance provides a direct measure of the system's complexity, we would expect that the variances would be proportional to one another, resulting in the AR

$$\langle V_2 \rangle \propto \langle V_1 \rangle^{h_1/h_2}. \tag{3.31}$$

This is precisely the logic adopted in Section 3.2.3 to obtain the metabolic rate AR from the data on BMR variances.

# 4 Allometry theories

Two distinct strategies dominate the many approaches adopted to ferret out the origins of ARs and for interpreting allometry parameters. One is based on a first-principles reductionistic approach, which starts by assuming forms for the underlying mechanisms and from those assumptions, deduce the implied algebraic form of the AR. Another method is phenomenological and involves statistical analysis, as well as, identifying patterns in the data. These patterns are then used to deduce the classical algebraic forms. A number of the statistical methods were reviewed in the previous chapter. The ARs stand out as empirical patterns that have withstood the test of time, whereas the same cannot be said for the models developed to explain how they come about. In this chapter we review some general principles, from which the underlying mechanisms, whether reductionist or statistical, have been identified and used to produce the ARs.

The search for *unifying principles* outside the physical sciences has systematically encountered barriers to establishing their existence. These barriers have often been traced to the underlying complexity of the phenomena of interest and ways of overcoming these barriers frequently rely on optimization approaches, as we demonstrate. Investigators in the social and life sciences have recently rediscovered Newton's 'principle of similitude' introduced in the *Principia* (II, Proposition 32), as a way of understanding complexity. Scaling and the principle of similitude have been present in the study of complex physical phenomena, since physics became a science of quantification. In modern times the principle of similitude has been nudged to the sidelines by renormalization group theory (RGT), which provides a formalism for determining how forces are transferred across multiple scale [162, 173, 380].

Part of the reason for exploring scaling and the RGT approach is that fractal geometry and fractal statistics are able to capture some of the 'regularity' observed in vast amounts of data in the life and social sciences, in addition to the physical sciences. The principle of similitude may be applied to fractal phenomena, even though mechanical forces may be ill-defined. The implementation of fractal geometry and RGT to study the architecture of physiological forms [25, 338, 365], interacting networks of chemical reactions [85, 126, 258] and the topology of ecological webs [50], over the past quarter century has lead to some remarkable insights in all these areas. In particular, the descriptive successes of investigations using fractal ideas [22, 201, 205, 214] suggests that complex networks with fractal architectures have an evolutionary advantage [340, 341, 344], as do fractal stochastic processes [345], both of which were discussed in Chapter 1.

In this chapter we show how the understanding of complexity barriers naturally leads to replacing the stochastic differential equations for the dynamics of large numbers of individuals by partial differential equations [354], to describe the dynamic behavior of ensembles of such individuals. The steady-state analytic solution to the PDF

DOI 10.1515/9783110535136-004

equations, enable us to explicitly evaluate the averages contributing to the empirical ARs and thereby derive the empirical AR from first principles [352].

## 4.1 Optimization principles

Optimization provides one of the better recipes for determining the dynamics that is consistent with what is empirically known about a phenomenon. It is through the specification of constraints that empirical information, about complex phenomena, is introduced into dynamic models. The biological sciences employ a number of different 'principles', such as energy minimization, the minimal use of materials, along with the maximization of efficiency and most recently, entropy generation minimization. So whether the extremum considered is a minimum or a maximum depends on one's purpose and the quantity being investigated. In addition to these familiar quantities from the physical and biological sciences, we are now sensitive to the need for a measure of complexity, as well as, a way to describe information flow within both physical and non-physical complex phenomena.

### 4.1.1 Energy minimization

Scaling relations in living networks result from the balance between various constraints. One such biological balance is that between the amount of energy available and the energy cost to carry out a biological function. Another technique that has been used both implicitly and explicitly in the derivation of biological ARs is modeling the transport of nutrients to various parts of an organism, through the venous and capillary networks, as well as, through the respiratory network.

The simplest model of fluid transport within physiological systems consists of pipes or tubes and the flow of fluid through them. Murray [228] considered a fluid with viscosity $v$, laminar flow $Q$, within a tube of length $l$ and radius $r$. The flow must overcome the vascular resistance producing a pressure difference $\Delta p$ along the length of the tube given by Poiseuille's Law

$$\Delta p = 8lvQ/\pi r^4. \tag{4.1}$$

A constraint on the flow is the cost of transporting fluid of cylindrical volume $V = \pi lr^2$ along the tube. Introducing the cost factor $c$, the total work done per unit time is given by

$$E = Q\Delta p + cV = \frac{8lvQ^2}{\pi r^4} + c\pi lr^2, \tag{4.2}$$

a quantity the living network of tubes makes as small as practical to carry out its function. Consequently, minimizing this expression with respect to the tube radius

$$\frac{\partial E}{\partial r} = -\frac{32lvQ^2}{\pi r^5} + 2c\pi lr = 0$$

yields the optimal flow:

$$Q = Cr^3, \tag{4.3}$$

with the constant given by

$$C = \sqrt{c\pi^2/16v}. \tag{4.4}$$

The cubic dependence of the flow rate on radius is known as Murray's law [331] and is indicative of the maximum efficiency of the flow through the tube.

Murray [229] subsequently extended his result to bifurcating networks such as occurs in bronchial airways. The flow from a parent vessel of radius $r_0$, divides into two daughter vessels of radii $r_1$ and $r_2$, such that the flow divides additively:

$$Q_0 = Q_1 + Q_2,$$

as had been observed some four centuries earlier by da Vinci. Therefore, inserting Murray's law into this expression yields da Vinci's equation with $\alpha = 3$ and in the case of equal radii in the daughter branches $r_1 = r_2$ yields the scaling relation

$$r_1 = 2^{-1/3} r_0. \tag{4.5}$$

Thus, the maximally efficient bifurcating network, in terms of energy transport cost, has radii decreasing from generation to generation as the cube root of two. As pointed out by Weibel [331], this last result is also known as Murray's law. However this law was actually first formulated by Hess [143] for blood vessels; subsequently, it should more properly be named the Hess–Murray law.

### 4.1.2 Optimal design

Rashevsky [262] introduced the *Principle of Optimal Design* (POD) in which the combination of the material used and the energy expended to achieve a prescribed function, is minimal. He applied POD to the basic problem of how the arterial network could branch in space in order to supply blood to every element of tissue within an organ. To address this problem, he used the model of a bifurcating branching network supplying blood to a restricted volume and reducing the total resistance to the flow of blood. His purpose was to determine the condition imposed by the requirement that the total resistance is minimum.

Here we assume the branching network is composed of $N$ generations from the point of entry (0) to the terminal branches ($N$). A typical tube at some intermediate

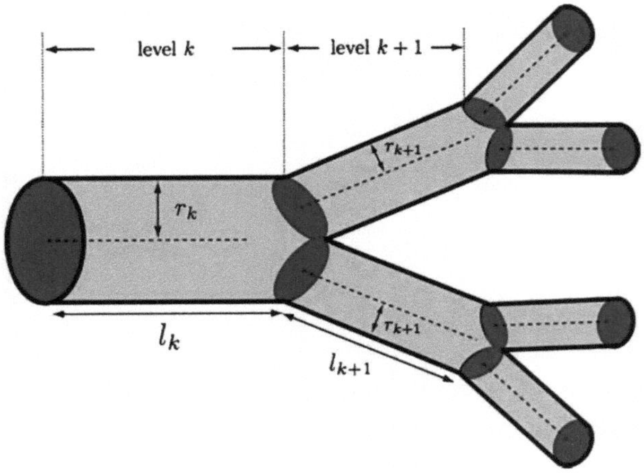


**Figure 4.1:** Sketch of a branching structure such as a blood vessel or bronchial airway with the parameters used in a bifurcating network model [262].

generation $k$ has length $l_k$, radius $r_k$, as depicted in Figure 4.1 for $q$ bifurcating network, and pressure drop across the length of the branch $\Delta p_k$. The volume flow rate in this generation of the tube $Q_k$ is expressed in terms of the flow velocity averaged over the cross sectional area $u_k$:

$$Q_k = \pi r_k^2 u_k. \tag{4.6}$$

Each tube branches into $n$ smaller tubes with the branching of the vessel occurring over some distance that is substantially smaller than the lengths of the tubes of either generation. Consequently, the total number of branches generated up to generation $k$ is $N_k = n^k$. The pressure difference at generation $k$, between the ends of a tube, is given by a suitably indexed version of Poiseuille's Law. The resistance $\Omega_k$ to the flow, at generation $k$, is given by the ratio of the pressure difference from equation (4.1) to the flow rate

$$\Omega_k = \frac{\Delta p_k}{Q_k} = \frac{8\nu l_k}{\pi r_k^4}. \tag{4.7}$$

The total resistance for a network branch with $m$ identical tubes in parallel is $1/m$ times the resistance of each individual tube. Thus, in this simplified case, we can write the total network resistance in terms of equation (4.7) as

$$\Omega_T = \frac{8\nu l_0}{\pi r_0^4} + \frac{8\nu}{\pi}\sum_{j=1}^{N}\frac{1}{N_j}\frac{l_j}{r_j^4}. \tag{4.8}$$

In order to minimize the resistance for a given mass, Rashevsky first expressed the initial radius $r_0$ in terms of the total mass of the network. The optimum radii for the different branches of the bifurcation network having the total mass $M$ are then determined such that the total resistance is a minimum

$$\frac{\partial \Omega_T}{\partial r_k} = -\frac{32\nu}{\pi}\left(\frac{l_0}{r_0^5}\frac{\partial r_0}{\partial r_k} + \frac{1}{N_k}\frac{l_k}{r_k^5}\right) = 0$$

and using [262]:

$$r_0 = \sqrt{\frac{1}{l_0}\left(\frac{M}{2\pi\alpha} - \sum_{j=1}^{N} N_k r_k^2 l_k\right)},$$

simplifies the minimization expression to

$$\frac{\partial \Omega_T}{\partial r_k} = \frac{32\nu}{\pi}\left(\frac{N_k r_k l_k}{r_0^6} - \frac{l_k}{N_k r_k^5}\right) = 0,$$

yielding the scaling rule

$$r_k = N_k^{-1/3} r_0. \tag{4.9}$$

The ratio of the radii between successive generations can consequently be expressed in terms of the ratio of the number of tributaries in successive generations

$$r_{k+1}/r_k = (N_k/N_{k+1})^{1/3},$$

so that inserting the number of branches at the $k^{\text{th}}$ generation $N_k = n^k$ yields

$$r_{k+1}/r_k = n^{-1/3}. \tag{4.10}$$

As observed earlier, this ratio corresponds to an exponential reduction in the branch radii across $k$ generations

$$r_k = r_0 e^{-k\lambda_n}; \quad \lambda_n = \frac{1}{3}\ln n. \tag{4.11}$$

Note the formal similarity of this ratio to Horton's Law, in which the ratio of numbers of river branches is independent of generation number.

Rashevsky considered the bifurcating case $n = 2$ where the ratio of radii reduces to

$$r_{k+1}/r_k = 2^{-1/3} = 0.794. \tag{4.12}$$

This is the classic 'cube law' branching discussed by Thompson [317], which he obtained using the 'principle of similitude'. The value $2^{-1/3}$ was also obtained by Weibel and Gomez [330] for the reduction in the diameter of bronchial airways for the first ten generations of the bronchial tree of the mammalian lung. However, they noted a sharp deviation away from this constant fractional reduction after the tenth generation, see Figure 4.2. The value $2^{-1/3}$ was again obtained by Wilson [381], who explained the proposed exponential decrease in the average radius of a bronchial tube, with generation number, by showing that this is the functional form for which a gas of given composition can be provided to the alveoli, with minimum metabolism, or entropy production in the respiratory musculature. He proposed minimum entropy production [115] as the design principle for biological networks to carry out a given function. We subsequently introduced the information equivalent of this design principle.

The deviation from classical (exponential) scaling above generation ten, shown in Figure 4.2, was eventually explained using an alternative model of the bronchial airways in terms of fractal statistics [338].

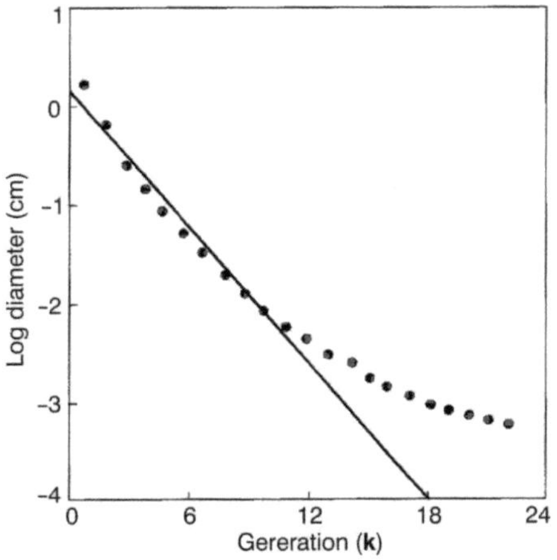


**Figure 4.2:** As the bronchial tree branches out, its tubes on average decrease in size. A theory consistent with the Principle of Similitude [317] predicts that their diameters should decrease by about the same ratio from one generation to the next; exponential reduction between successive generations. This semilog graph shows measurements from Weibel and Gomez [330] for 23 generations of bronchial tubes in the human lung. The prediction is a straight line that fits the anatomic data (dots) until about the tenth generation, after which they deviate systematically from an exponential decline. (Adapted from [339] with permission.)

### 4.1.3 Why fractals?

In Chapter 1 we discussed the evolutionary advantage of fractal design. One consequence of the scale-free nature of fractal processes is their relative insensitivity to random errors. For example, during morphogenesis, chemistry controls the growth of an organism. There is a certain variability tolerance that is built into the biochemical processes that control growth, without which the organism could not survive. In that earlier discussion we determined that a fractal growth process was relatively insensitive to random errors and therefore provided a survival advantage to an organism that adopts this strategy during macro-evolution. It seems that once nature finds such a strategy she adopts it everywhere we look. Or is it once we acquire a hammer then everything appears to be a nail?

There has never been universal acceptance of the notion that systems in biology and sociology adopt optimization procedures, or that they, in fact, should. The engineering criteria of optimal design has always appeared, to some, as an overly restrictive constraint on living organisms and organizations. Living systems survive by anticipating and adapting to a broad spectrum of environmental changes, which would appear to be incompatible with having a single optimization strategy. A more reasonable optimization strategy would incorporate a broad spectrum of responses to manage the spectrum of possible perturbations from the environment, or from another system with which it interfaces.

In his excellent book *Symmorphosis* [331], Weibel introduces the radical principle of Symmorphosis. This new principle postulates:

> ...the precise tuning of all structural elements in an organism to each other and to the overall functional demand.

The interface between two such 'structural elements' or subnetworks cannot be more efficient as a transfer surface than that of two interleaving fractals. This has appeared in the context of information transfer between two complex networks, which has been used as a proof of the Principle of Complexity Management (PCM). In this latter situation the relative complexity of two interacting networks determines the direction of information flow and the efficiency of that information exchange [11, 347].

## 4.2 Scaling and allometry

Barenblatt and Monin [21] proposed that metabolic scaling might be a consequence of the fractal nature of biology, but they did not provide a mathematical model for its description. This shortcoming has been overcome by a number of investigators who have devised numerous fractal scaling models to describe AR in a variety of contexts [338, 365].

### 4.2.1 Elastic similarity model

One of the first models to analytically predict the allometry exponent 3/4 was constructed by McMahon [209, 210] and was introduced before the coining of the term fractal. His argument rests on quantifying the observation that the weight of a column increases more rapidly with size, than does its strength. Moreover, as discussed by Schmidt-Nielsen [288], in his book *Scaling, Why is Animal Size so Important*, if a column is tall and slender, it can fail due to elastic buckling in which small lateral displacements exceed the elastic restoring forces. For a sufficiently slender column with Young's elastic modulus $E$ for the material, density $\rho$ and diameter $d$, the critical length $l_{cr}$ is

$$l_{cr} = k(E/\rho)^{1/3}d^{2/3} \tag{4.13}$$

and $k$ is a known constant. The elastic criteria of McMahon is therefore given by

$$l_{cr}^3 \propto d^2. \tag{4.14}$$

This supports the intuitive argument of Galileo mentioned earlier that the weight increases as the cube of a characteristic scale, whereas the strength increases only as the square.

The interest in McMahon's model would have waned if he had stopped with the observation of equation (4.14). However, he went on to develop the idea that the weight of the column is proportional to the product of the density, length and cross-sectional area,

$$Mg \propto \rho d^{2/3}\pi d^2,$$

where the length has been replaced using the elastic criteria. The diameter of the column is therefore proportional to its total mass raised to a curious power

$$d \propto M^{3/8}. \tag{4.15}$$

Schmidt-Nielsen [288] points out that McMahon presented empirical data supporting the elastic similarity model [210]. One such data set consisted of 3 000 Holstein cattle and the measure of size was the average height at the withers, which replaces the critical length in the elastic similarity model,

$$H \propto M^{0.24},$$

with the empirical exponent of 0.24 compared with the predicted exponent of 0.25.

The implications of the scaling between the cylinder diameter and mass for allometry was discussed by Calder [57], which enabled him to determine the metabolic allometry exponent. In his argument he used the *Symmorphosis Hypothesis* of Taylor and Weibel [316]:

> ...is regulated to satisfy but not exceed the requirements of the functional system.

The function referred to in the Symmorphosis hypothesis in the present case is locomotion and the regulation is determined by the metabolic rate. First off, we recognize that locomotion requires the contraction of muscles. During contraction, muscles exert a force that increases with the cross-sectional area of the muscle, as mentioned above. The power output of the muscle is the work done per unit time and may be equated with the metabolic rate, the product of the force generated by the muscle and the velocity $u$ of the shortening of the muscle

$$B \propto d^2 u.$$

The velocity of the shortening of the muscle appears to be a size-independent constant from species to species [146], so that using the elastic similarity relation of the diameter given by equation (4.15) to mass yields the BMR

$$B \propto M^{3/4}. \tag{4.16}$$

Consequently, the allometry exponent 3/4 in McMahon's elastic similarity model is required to maintain the flow of energy to the working muscles and is consistent with the Principle of Symmorphosis.

Versions of the above argument, given by both Calder [57] and Schmidt-Nielsen [288], seem to explain the value of the allometry exponent for warm blooded animals. Dodds et al. [89] critique McMahon's model by noting that there is no compelling reason to believe that the power output of muscles should be the dominant factor in the scaling of BMR. Moreover, Savage et al. [282] point out that while the elastic similarity model might apply to the bones of mammals, or the trunks of trees that have adapted to gravitational forces, it is doubtful that it is applicable to aquatic or unicellular organisms that also appear to display an allometry exponent of 3/4 [141].

### 4.2.2 WBE model

A model of metabolic ARs that takes into account the scaling of physiological networks was proposed nearly two decade ago by the physicist, biologist and ecologist, West, Brown and Enquist, respectively [365]. This model of nutrient transport within a hierarchal network, in which vessels become narrower, shorter and more numerous, between successive generations, proceeding from the distal to the terminal generation, is reminiscent of the river branchings discussed earlier. The hydrodynamic scaling in the transport network is entailed by three fundamental assumptions: (1) The entire volume of an organism being supplied resources is crowded with a space-filling branching network of tubes. (2) The tube properties at the terminus of the network are size-invariant. (3) The energy required to transport resources using this network is minimal, that is, the hydrodynamic resistance of the network is minimized. They assert that this model describes the origin of universal scaling laws in biology [366, 368], resulting in a metabolic AR with $b = 3/4$. However, we recall at the start of this discussion that the existence of a fixed empirical exponent for metabolic AR has been questioned by numerous investigators [89, 145, 174, 175, 198] and is inconsistent with the covariation function discussed in the previous chapter.

We note that the original hydrodynamic argument [365] was replaced a couple of years later in their sequel [367], with a more general geometric scaling argument for hierarchical networks. We refer to the latter as the WBE model. The WBE analysis shows that under a reasonable set of assumptions, the AR, between the area of an organism and its mass has an allometry exponent $b = 3/4$. This is one of those models that is so elegant in both its generality and simplicity, that it ought to be true. Unfortunately for the universality argument, data show that the allometry exponent $b$ and the allometry coefficient $a$ covary [118, 119], so that $b$ cannot be restricted to a single value.

In Chapter 3 we used the PEC to explain the dependence of the allometry exponent on the allometry coefficient, which argues further against universality. The skeptical reader may then question the reason for including a discussion of the WBE model in this essay. The reason is that the logic, underlying the scaling argument, is still applicable, but to the PDF and not the variables themselves, as we subsequently explain. Moreover, the final word in this area of research has not been written, so that it would be a disservice to the reader not to, at least, sketch out the alternative theories.

The WBE model is based on a fundamental assumption: nature selects organisms to maximize fitness, which it does by maximizing the rate of uptake of energy and material from the environment for allocation to survival and reproduction. This assumption entails the geometrical scaling of the effective surface area of the organism $\mathcal{A}$, as measured by the BMR $B$, through which nutrients and energy are exchanged, is maximum. This conjecture is equivalent to the statement of proportionality between the two functionality variables

$$B \propto \mathcal{A}. \tag{4.17}$$

An additional assumption made for the WBE model involves minimizing the time and resistance for internal transport of resources, by minimizing some characteristic length scale of the hierarchical network.

Following the discussion of WBE, consider an organism whose effective surface area depends on a fixed terminal scale length $l_0$ for the biological network, in addition to various other length scales $l_i$ that parameterize the network's fractal-like structure. This allows us to write the effective exchange area as

$$\mathcal{A}(l_0, l_1, l_2, \ldots) = l_1^2 \Phi\left(\frac{l_0}{l_1}, \frac{l_2}{l_1}, \ldots\right), \tag{4.18}$$

where $\Phi$ is a dimensionless function of the ratio of lengths. Introducing an arbitrary scale transformation on the network

$$l_i \to l_i' = \lambda l_i; \quad i = 1, 2, \ldots \tag{4.19}$$

allows us to replace equation (4.18) with

$$\begin{aligned} \mathcal{A} \to \mathcal{A}' &= \mathcal{A}(l_0, \lambda l_1, \lambda l_2, \ldots) \\ &= \lambda^2 l_1^2 \Phi\left(\frac{l_0}{\lambda l_1}, \frac{l_2}{l_1}, \ldots\right), \end{aligned} \tag{4.20}$$

since the terminal length $l_0$ is fixed and therefore does not scale. Not knowing the $\lambda$-dependence of the RHS of equation (4.20), WBE make the educated guess that it ought to reflect the hierarchical character of the fractal-like organization and assume

$$\Phi\left(\frac{l_0}{\lambda l_1}, \frac{l_2}{l_1}, \ldots\right) = \lambda^{\varepsilon_a} \Phi\left(\frac{l_0}{l_1}, \frac{l_2}{l_1}, \ldots\right) \tag{4.21}$$

and, up to this point, the exponent $\varepsilon_a$ is left unspecified, resulting in

$$\begin{aligned} \mathcal{A} \to \mathcal{A}' &= \mathcal{A}(l_0, \lambda l_1, \lambda l_2, \ldots) \\ &= \lambda^{2+\varepsilon_a} \mathcal{A}(l_0, l_1, l_2, \ldots). \end{aligned} \tag{4.22}$$

WBE emphasize that the fixed terminal length $l_0$ modifies the Euclidean scaling of area $\lambda^2$ and they introduce the fractal dimension for the effective area

$$D_a \equiv 2 + \varepsilon_a. \tag{4.23}$$

This assumption enabled them to interpret the parameter they introduced to lie in the range $0 \le \varepsilon_a \le 1$. The lower extreme defines the traditional Euclidean area of the organism, the upper extreme defines a volume filling structure of "maximum fractality" in which the effective area scales like the Euclidean volume.

They go on to consider the effective volume $\mathcal{V}$ associated with the effective area $\mathcal{A}$:

$$\mathcal{V}(l_0, l_1, l_2, \ldots) = l_1^3 \Theta\left(\frac{l_0}{l_1}, \frac{l_2}{l_1}, \ldots\right) \tag{4.24}$$

where $\Theta(\cdot)$ is a dimensionless function of the ratios of lengths. Equation (4.24) represents the volume of biologically active material within the organism and is not necessarily the same as the Euclidean volume. In parallel with the treatment of the effective area, they introduce a scaling parameter $\lambda$ such that

$$\Theta\left(\frac{l_0}{\lambda l_1}, \frac{l_2}{l_1}, \ldots\right) = \lambda^{\varepsilon_\upsilon} \Theta\left(\frac{l_0}{l_1}, \frac{l_2}{l_1}, \ldots\right) \tag{4.25}$$

resulting in the scaling behavior

$$\begin{aligned} \mathcal{V} \to \mathcal{V}' &= \mathcal{V}(l_0, \lambda l_1, \lambda l_2, \ldots) \\ &= \lambda^{3+\varepsilon_\upsilon} \mathcal{V}(l_0, l_1, l_2, \ldots). \end{aligned} \tag{4.26}$$

Here again the "arbitrary" parameter $\lambda$ allows us to introduce the effective volume fractal dimension

$$D_\upsilon = 3 + \varepsilon_\upsilon, \tag{4.27}$$

so that $0 \le \varepsilon_\upsilon \le 1$.

Combining equations (4.26) and (4.22) allows us to relate the effective area and volume in the following way:

$$\mathcal{A} \propto \mathcal{V}^{\frac{D_a}{D_\upsilon}}. \tag{4.28}$$

Note that this is a generalization of the scaling arguments made over two centuries ago to balance the heat generated within a volume of metabolic reactions and dissipated across the encompassing surface area, thereby yielding the 2/3-law. One might now refer to the above scaling argument as the $D_a/D_\upsilon$-law for metabolic fractal organisms.

The final argument in their logical chain is that the scaling given by equation (4.18) can be carried out once more on the linear scale, just as was done for area and volume. Identifying the volume as the product of this linear scale with that of the effective area allows for the volume parameter to be written as

$$\varepsilon_v = \varepsilon_a + \varepsilon_l. \tag{4.29}$$

Consequently, they assume a uniform constant density that implies the volume is proportional to the mass, resulting in

$$\mathcal{A} \propto \mathcal{V}^{\frac{D_a}{D_v}} \propto M^{\frac{2+\varepsilon_a}{3+\varepsilon_a+\varepsilon_l}}, \tag{4.30}$$

which from equation (4.17) yields the schematic AR

$$B = aM^{\frac{2+\varepsilon_a}{3+\varepsilon_a+\varepsilon_l}}, \tag{4.31}$$

with the allometry exponent given by

$$b = \frac{2+\varepsilon_a}{3+\varepsilon_a+\varepsilon_l}. \tag{4.32}$$

Finally, in order for their conjecture, that the effective area is maximum, to be realized, it is necessary that the allometry exponent also be maximum. The maximum occurs at $\varepsilon_a = 1$, where the effective area becomes volume filling and $\varepsilon_l = 0$, where the transport time within the network is minimized, resulting in $b = 3/4$. In the WBE scaling model the allometry exponent has the universal value 3/4, regardless of the details of the branching architecture and independently of the dynamics governing the metabolic processes and the transport of nutrients.

It is worth mentioning that Weibel [332] presents a simple and compelling argument on the limitations of the WBE model in terms of transitioning from BMR to the maximal metabolic rate (MMR) induced by exercise. The AR for MMR has an exponent $b = 0.86$ rather than 3/4, so that a different approach to determining the exponent is needed. Painter [239] demonstrates that the empirical allometry exponent for MMR can be obtained in the manner pioneered by the nutrient transport model [365], by using the Hess–Murray Law for the scaling of branch sizes between levels.

Weibel [332] argues further that a single cause for the power function arising from a fractal network is not as reasonable as a model involving multiple causes, see also Agutter and Tuszynski [2]. Darveau et al. [81] proposed such a model, recognizing that the metabolic rate is a complex property resulting from a combination of functions. West et al. [370] and Banavar et al. [18] demonstrate that the mathematics in the distributed control model of Darveau et al. [81] is fundamentally flawed mathematically. In their reply, Darveau et al. [82] do not contest the mathematical criticism, instead they point out consistency of the multiple-cause model of metabolic scaling with what is known from biochemical [303] and physiological [157] analysis of metabolic control. The notion of distributed control remains an attractive alternative to the single cause models of metabolic AR. A mathematically rigorous development of AR with fractal responses from multiple causes was recently given by Vlad et al. [324] in a general context. This latter approach may answer the many formal questions posed by these critics.

### 4.2.3 The optimum can be dangerous

A question of interest is whether optimal design criteria are ever realized in nature and whether they are even desirable? The WBE model suggests that the AR exponent value of 3/4 is proof of the optimally of fractal design of physiological networks. However, the controversy over the empirical value of the allometry exponent calls the applicability of the 'proof' into question, which is to say, although the mathematics cannot be faulted, its applicability to a wide range of physiological networks remains doubtful.

Another distinct application of the fractal design principle was made to the mammalian lung by West et al. [338], who established that the average diameter of a bronchial tube, as a function of generation number, is described by a modulated IPL. This discussion regarding the mammalian lung and whether the bronchial tree is optimal has been investigated by Mauroy et al. [208]. The latter authors maintain that the mammalian bronchial tree is a good example of an efficient distribution network, with an approximate fractal structure [230, 338]. They state that physical optimization is critical in that small variations in the geometry can induce large variations in the net air flux and consequently, optimality cannot be a sufficient criterion for physiologic design of the bronchial tree. The slight deviations observed in the parameters presumed to be optimized are a manifestation of a safety factor that has been incorporated into the design and hence into the capacity for regulating airway caliber.

In the present context the size ratio $h$ of successive airway segments are homothetic with $h = 2^{-1/3} \approx 0.79$, as discussed earlier. Homothetic scaling means that the lengths and diameters have the same ratios, between successive generations. Using the resistance minimization argument for the bronchial network Mauroy et al. [208] show that the 'best' bronchial tree is fractal, with constant reduction factor given by the Hess–Murray Law. Do the data support this optimal value? If not, what does that imply about the efficiency of bronchial airways?

The fractal dimension for a bronchial airway is $D = -\ln 2/\ln h$ so that the Hess–Murray Law implies $D = 3$, whereas $h > 0.79$ implies $D > 3$. In the human lung it is found that the homothety ratio is $h \approx 0.85$ [332] and the bronchial network is therefore not optimized: its volume is too large and its overall resistance is too small. Mauroy et al. [208] emphasize that this deviation from optimal is, in fact, a safety margin for breathing, with respect to possible bronchial constrictions.

Sapoval [281] has argued that without regulation of the airway caliber [259], there would be a multifractal spatial distribution of air within the lungs, resulting in strongly non-uniform ventilation, with some regions of the lung being poorly fed with fresh air. Expanding on this theme, using inhomogeneity of the homothety ratio, Mauroy et al. [208] show how the optimal network is dangerously sensitive to physiological variability and consequently effective design of the bronchial tree must incorporate more than just physical optimality. This argument has clear implications for other scaling networks as well. One implication is the need to replace scaling, given by fractal geometry, with fractal statistics, to which we now turn our attention.

## 4.3 Stochastic differential equations

Fractal processes are rich in scales, with no one scale being dominant, either in the geometry, statistics or dynamics. Thus, information in fractal phenomena is coupled across multiple scales, manifesting long-time memory, as for example, observed in the architecture of the mammalian lung [230, 331, 338]; manifest in the long-range correlations in human gait [138, 344] and measured in the human cardiovascular network [243], all of which are reviewed in West [346]. These phenomenological characterizations of fractal time series relate back to the observation made earlier that if $X(t)$ and $Y(t)$ are each stochastic variables, then $Y(X)$ is a function of its random argument as well. Consequently, the general approach to determining the scaling relation between such variables is through their PDFs, obtained from solutions to the phase space equations, equivalent to the stochastic differential equations (SDE).

### 4.3.1 Stochastic dynamics

One strategy for treating random dynamic processes uses generalizations of the dynamic equations, originally constructed by Langevin [178]. He introduced uncertainty into the force law governing the motion of a Brownian particle immersed in a fluid by introducing a random force into the equation of motion for the Brownian particle. The random force arises from the instantaneous imbalance in the collisons of the ambient particles with the surface of the larger and more massive Brownian particle. This picture has matured into the random force characterizing the interaction of a quantity of interest with its uncertain environment.

A second strategy for the treatment of uncertainty is based on the phase space evolution of a PDF, using the Fokker–Planck equation (FPE) or even more general phase space equations (PSEs). The conditions under which these two strategies, the phase space equation for the PDF and the single trajectory dynamics for the Brownian particle, are equivalent have been shown in a number of places, see, for example, Lindenberg and West [184].

Consider the dynamics of a simple exponential relaxation process $Z(t)$, which is disrupted by a random force $\xi(t)$:

$$\frac{dZ(t)}{dt} = -\lambda Z(t) + \xi(t). \tag{4.33}$$

We restrict our remarks to one dimension in order to not obscure the discussion, with possibly confusing notation. In the familiar case of the Brownian motion of a heavy particle immersed in a fluid of lighter particles, the dynamic variable $Z(t)$ is the velocity of the heavy particle of unit mass. The dissipative term is produced by Stokes drag on the heavy particle and is proportional to the velocity, through a friction rate $\lambda$. The random force $\xi(t)$ is due to the imbalance in the buffeting on the heavy particle's surface by the lighter particles of the ambient fluid. In a more general context the random force is the influence of the unknown and unknowable environment on the dynamics of the phenomenon of interest and $Z(t)$ is the dynamic observable.

We assume, based on the central limit theorem, that for a Newtonian fluid, the statistics of the random force are Gaussian, with a delta correlated correlation of strength $D$:

$$\langle \xi(t)\xi(t') \rangle = 2D\delta(t-t'), \tag{4.34}$$

where the brackets denote an average over an ensemble of realizations of the random force. In this case the statistical fluctuations are modeled by a Wiener process $dW(t) = \xi(t)dt$. Note that in complex fluids such as blood, or biofluids containing micro-organisms, such simplifying assumptions break down and must be replaced with properties more compatible with experimental observations. The fluctuations can have memory, in which case equation (4.34) no longer holds, or the statistics can be non-Gaussian, in which case the statistical process is no longer Wiener, or both, for these more complex phenomena. Such cases are discussed subsequently in an allometry context.

The solution to the simple stochastic rate equation (4.33) is

$$Z(t) = e^{-\lambda t}Z(0) + \int_0^t e^{-\lambda(t-t')}\xi(t')dt'. \tag{4.35}$$

The average solution, is obtained by averaging equation (4.35) over an ensemble of realizations of the zero-centered random force, to obtain

$$\langle Z(t) \rangle = e^{-\lambda t}Z(0), \tag{4.36}$$

with variance given by

$$\langle Z(t)^2 \rangle - \langle Z(t) \rangle^2 = \frac{D}{\lambda}(1 - e^{-2\lambda t}). \tag{4.37}$$

This is the well known result for the Wang–Uhlenbeck process [326] given by equation (4.33). Note that in classical diffusion the coefficient of friction is assumed to be negligible

$$\lim_{\lambda\to 0}[\langle Z(t)^2 \rangle - \langle Z(t) \rangle^2] = 2Dt, \tag{4.38}$$

resulting in a variance that grows linearly in time. However, for finite dissipation the average decays exponentially in time and the variance approaches a constant value, given by the ratio of the diffusion coefficient and the dissipation rate. With $Z$ identified as the velocity of a Brownian particle, the variance would be proportional to the temperature $T$ of the surrounding fluid, resulting in the Einstein fluctuation–dissipation relation

$$D/\lambda = k_B T, \tag{4.39}$$

see, for example, Fürth [108].

It is often useful to consider the characteristic function, defined by the Fourier transform of the PDF

$$\phi(k,t|z_0) \equiv \mathcal{FT}\{P(z,t|z_0);k\} = \int_{-\infty}^{\infty} dz e^{ikz} P(z,t|z_0), \tag{4.40}$$

where the quantity $P(z,t|z_0,t_0)dz$ is the probability that the dynamic variable lies in the phase space interval $(z+dz,z)$ at time $t$ conditional on $Z(t_0)=z_0$ at the initial time $t=t_0$. The characteristic function can also be written

$$\phi(k,t|z_0) = \langle e^{ikZ(t)} \rangle, \tag{4.41}$$

where $Z(t)$ is the solution to the SDE. In equation (4.41) the bracket refers an average over an ensemble of realizations of the statistical fluctuations in the solution to the SDE. Consequently, by inserting equation (4.35) into (4.41) and using the definition of the mean equation (4.36) and variance equation (4.37), we obtain for the characteristic function, assuming $t_0 = 0$:

$$\phi(k,t|z_0) = e^{ik\langle z;t\rangle} e^{-\sigma^2(t)k^2}. \tag{4.42}$$

We have used the expansion

$$\begin{aligned}\langle e^{i\eta}\rangle &= \sum_{n=0}^{\infty} \frac{\langle (i\eta)^n\rangle}{n!} \\ &= \sum_{n=0}^{\infty} \frac{(-1)^n \langle \eta^2\rangle^n}{n!} = e^{-\langle \eta^2\rangle}\end{aligned}$$

for a zero-centered Gaussian variable $\eta$, to obtain the characteristic function equation (4.42). The PDF solution is obtained by taking the inverse Fourier transform of the characteristic function given by equation (4.42):

$$P(z,t|z_0) \equiv \mathcal{FT}^{-1}\{\phi(k,t|z_0);z\} = \int_{-\infty}^{\infty} \frac{dk}{2\pi} e^{-ikz} \phi(k,t|z_0). \tag{4.43}$$

Inserting equation (4.42) into (4.43) and carrying out the inverse Fourier transform, results in the time-dependent Gaussian PDF centered on the average value

$$P(z,t|z_0) = \frac{1}{\sqrt{4\pi\sigma^2(t)}} \exp\left[-\frac{(z-\langle z;t\rangle)^2}{2\sigma^2(t)}\right], \tag{4.44}$$

with average value

$$\langle z;t\rangle = z_0 e^{-\lambda t}, \tag{4.45}$$

which by construction is in agreement with equation (4.36) and variance

$$\sigma^2(t) = \frac{D}{\lambda}(1 - e^{-2\lambda t}), \tag{4.46}$$

which also, by construction, is in agreement with equation (4.37). Thus, the characteristic function contains complete information regarding the stochastic dynamic process.

As seen in the discussion of allometry fluctuations, multiplicative and not additive random processes can be the more important. Multiplicative fluctuations can be addressed by examining the FPE associated with a linear stochastic dissipation parameter [184, 286]. Consider the nonlinear rate equation with multiplicative fluctuations

$$\frac{dQ}{dt} = -\lambda Q \ln Q + \xi(t)Q \tag{4.47}$$

for which the logarithmic transformation $Z = \ln Q$ yields the rate equation, with additive fluctuations $\xi(t)$ given by equation (4.33). It is worth noting that this is exactly the kind of logarithmic transformation done on the data to determine the allometry parameters. The solution to the SDE with multiplication fluctuations may be obtained using the solution to the SDE with additive fluctuations, the transformation of variables, and the conservation of probability:

$$P(q,t|q_0)dq = P(z,t|z_0)dz. \tag{4.48}$$

Inserting equation (4.44) into (4.48) yields the log-normal PDF

$$P(q,t|q_0) = \frac{1}{q\sqrt{4\pi\sigma^2(t)}} \exp\left[-\frac{(\ln q - \langle \ln q;t\rangle)^2}{2\sigma^2(t)}\right]. \tag{4.49}$$

Note that the log-normal PDF has an IPL asymptotically in $q$, for more details on log-normal processes, see Goel et al. [125].

This example was provided to give some indication of how the PDF can change, in going from additive to multiplicative fluctuations in the SDE. But the characteristic function method may not be the most efficient in dealing with SDE models of allometry, as we shall see.

### 4.3.2 Ontogenetic growth model

We now turn our attention to the dynamics in the background of the static ARs. At the turn of this century, West et al. [369] used the conservation of energy to deduce an ontogenetic growth model (OGM) for a single organism. The OGM equations were subsequently generalized to encompass interspecies growth, as well [227]. The generalized OGM is of the same form as one introduced half a century earlier by the champion of *Generalized Systems Theory,* von Bertalanffy [325]. The energy conservation equation interrelates the resting BMR denoted by $B$ and the TBM denoted by $M$ for a single organism:

$$E_m \frac{dM}{dt} = B - B_m M. \tag{4.50}$$

Moses et al. [227] explain that the BMR of the entire organism $B$ is the rate of metabolic energy generated at time $t$; $E_m$ is the metabolic energy required to create a unit biomass; and $B_m$ is the metabolic rate required to maintain an existing unit of biomass.

Following West and West [354] we insert the intraspecies AR for the BMR into equation (4.50) and using the notation introduced in Chapter 3 for the TBM of the $i^{\text{th}}$ individual of the $j^{\text{th}}$ species, obtain the OGM equations

$$\frac{dM_{ij}}{dt} = \alpha M_{ij}^b - \beta M_{ij}, \tag{4.51}$$

where the coefficients in equation (4.51) are the rate of creation of a unit biomass $\beta = B_m/E_m$ and the allometry coefficient normalized to the metabolic energy required to create a unit biomass $\alpha = a/E_m$. Moses et al. [227] relate the coefficients in equation (4.51) to those in the original OGM [369], where it was shown that the imbalance in the terms of the equation ultimately result in a steady-state solution satisfying

$$\frac{dM_{ij}}{dt} = 0. \tag{4.52}$$

This steady-state solution has a scaling form giving rise to a universal curve for $b = 3/4$ that is well fit to data sets for mammals, birds, fish, molluscs and plants. Banavar et al. [17] obtain an equivalent universal curve and fit the same data for 13 organisms equally well, for both $b = 3/4$ and $2/3$.

As Banavar et al. [17] point out, a more general rate equation was first considered by von Bertalanffy [325], who postulated a simple nonlinear rate equation to describe the growth of TBM $M$ of the form

$$\frac{dM}{dt} = A_\eta M^\eta - C_\gamma M^\gamma, \tag{4.53}$$

at time $t$, where $\eta$ and $\gamma$ are unspecified exponents, and $A_\eta$ and $C_\gamma$ are positives constants. The solution to this equation was studied by von Bertalanffy [325] with $\eta = 2/3$ and $\gamma = 1$. Equation (4.53) is the OGM solved by West et al. [369] to obtain a universal curve, which they fit to data using $\eta = 3/4$ and $\gamma = 1$. Banavar et al. [17] obtained an equivalent universal curve and fit the same data for both $\eta = 2/3, 3/4$ and $\gamma = 1$.

Here we follow the discussion of Banavar et al. [17] to solve the OGM equation having the scaling form

$$\frac{dM}{dt} = M^\eta f(M/M_0), \tag{4.54}$$

where $M_0$ is chosen to be the maximum TBM that solves the stationary equation

$$dM/dt = 0. \tag{4.55}$$

Rearranging the terms in equation (4.54) and introducing the dimensionless variable

$$r = (M/M_0)^{1-\eta}, \tag{4.56}$$

we can reduce the number of scaling parameters and write for $\gamma = 1$:

$$\frac{dr(t)}{dt} = \lambda_\eta[1 - r(t)], \tag{4.57}$$

with the rate of growth given by

$$\lambda_\eta \equiv \frac{(1-\eta)A_\eta}{M_0^{1-\eta}} \quad \text{and} \quad A_\eta = C_1 M_0^{1-\eta}. \tag{4.58}$$

The initial condition has the birth mass $M = m_0$ at $t = 0$, so that the solution to the OGM equation can be written as

$$-\ln[1 - r(t)] + \ln[1 - r(0)] = \lambda_\eta t, \tag{4.59}$$

with the initial dimensionless mass given by

$$r(0) = (m_0/M_0)^{1-\eta}. \tag{4.60}$$

Consequently, the solution to equation (4.57) can be expressed entirely in terms of dimensionless variables as

$$r(t) = 1 - [1 - r(0)]e^{-\lambda_\eta t}, \tag{4.61}$$

or in terms of the observables

$$\left[\frac{M(t)}{M_0}\right]^{1-\eta} = 1 - \left[1 - \left(\frac{m_0}{M_0}\right)^{1-\eta}\right]e^{-\lambda_\eta t}. \tag{4.62}$$

The plot of the multiple data sets collapsed onto a single universal curve in terms of the dimensionless mass $r$ versus the dimensionless time $\tau$:

$$\tau = \lambda_\eta t - \ln[1 - r(0)], \tag{4.63}$$

resulting in the universal solution

$$r(\tau) = 1 - e^{-\tau}. \tag{4.64}$$

The universal solution is depicted in Figure 4.3 and the data are fit equally well with $\eta = 2/3$ or $\eta = 3/4$ [17].

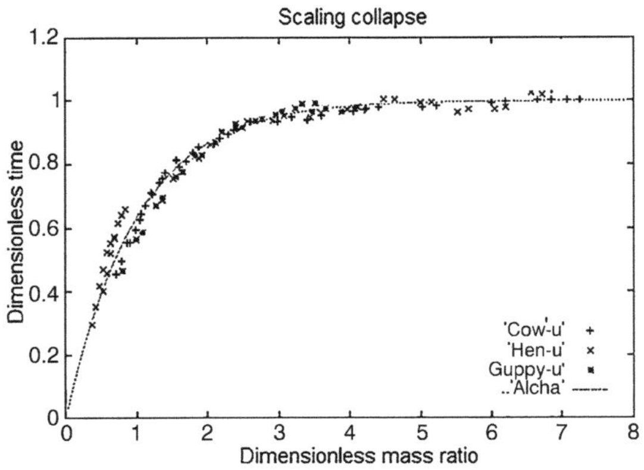


**Figure 4.3:** Growth curves for cow; guppy; and chicken. Best fit obtained by West et al. [369] (that is, $\eta = 3/4$); solid curve, plot of equation (4.64), with $\eta = 2/3$ and the values of $M$ and $\gamma$ as obtained in [369]. The universal growth curve (equation (4.64); dotted line) is derived from data from the three species (+, cow; ×, chicken; *, guppy) for both $\eta = 3/4$ and $\eta = 2/3$. (Adapted from [17] with permission.)

### 4.3.3 Growth of cities

Bettencourt et al. [35] use reasoning parallel to that developed in OGM to construct an equation for urban growth. They were able to do this after they established that, using the population $N(t)$ as the measure of city size and $Y(t)$ denoting material resources, or social activity, that there exists an urban AR:

$$Y(t) = aN(t)^b. \tag{4.65}$$

In the urban context there is no one accumulation point for the allometry exponent. In fact, depending on the functionality chosen, the allometry exponent falls into three categories: $b < 1$ (sublinear), associated with material quantities in the infrastructure; $b \approx 1$ (linear), associated with the needs of the individual, such as having a job or a house; $b > 1$ (superlinear), associated with societal things, such as information, innovation and wealth. An example of the superlinear case for wages as a function of city size was depicted in Figure 2.15.

These authors make the following observations regarding their data [35]:

> The most striking feature of the data is perhaps the many urban indicators that scale superlinearly ($b > 1$). These indicators reflect unique social characteristics with no equivalent in biology and are the quantitative expression that knowledge spillovers drive growth..., that such spillovers in turn drive urban agglomeration..., and that larger cities are associated with higher levels of productivity...Wages, income, gross domestic product, bank deposits, as well as rates of invention, measured by new patents and employment in creative sectors...all scale superlinearly with city size.

The superlinear property of urban ARs result in the rate of resource consumption scaling as $N^{b-1}$. This scaling of the rate supports the interpretation that urban processes, driven by innovation and wealth creation, produces the increase in the pace of urban life observed in the data [35]. We commented on the existence of this increased pace previously, but not on its cause.

In the urban growth model (UGM) of Bettencourt et al. [35] the resources $Y(t)$ are used for both growth and maintenance and are, in direct analogy with OGM, assumed to satisfy the equation

$$Y(t) = RN(t) + E\frac{dN(t)}{dt}, \tag{4.66}$$

where $R$ is the rate necessary to maintain an individual and $E$ the quantity necessary to add a new individual to the population. Substituting equation (4.66) into the urban AR given by equation (4.65) leads to the nonlinear rate equation

$$\frac{dN(t)}{dt} = \frac{a}{E}N(t)^b - \frac{R}{E}N(t), \tag{4.67}$$

which is of the general form of the von Bertalanffy [325] rate equation discussed in the last section. But the fact that the allometry exponent is not restricted to sublinear values, but has three domains, has some interesting consequences.

The three domains of the solution to the UGM rate equation are as follows. For $b = 1$ the solution is exponential, with the growth rate $(a-R)/E$. The $b < 1$ solution coincides with that discussed in the previous section, with its remarkable scaling collapse onto a single sigmoidal growth curve asymptotically reaching a finite population size. The solution in the superlinear domain $b > 1$, where growth is driven by innovation and wealth creation, differs markedly from the other two domains. First of all, for certain parameter values there is unbounded growth, which becomes singular at a critical time

$$t_c = \frac{E}{(b-1)R}\ln\left[1 - \frac{R}{a}N(0)^{1-b}\right]. \tag{4.68}$$

As Bettencourt et al. point out, this singularity can never be reached due to finite resources. However, they go on to emphasize that if growth is left unchecked, transition to a phase that reverses the sign within the logarithm occurs, triggering stagnation and ultimate collapse of the population.

The singularity and subsequent population collapse are therefore preventable, or at least postponable. As the city approaches such a singularity, a new critical time can be introduced to maintain growth and shift the critical time farther into the future. They explain that the city response must be "innovative" to ensure the dynamics remain in the "wealth and knowledge creation" phase, in which the sign of the term within the logarithm remains positive. Each reset of the critical time, by means of a creative response, initiates a new growth cycle, with the multiplicity of cycles extending the vitality of the city for longer periods. In this way they not only explain both the collapse of societies in the past, but they also offer hope for the stability of innovative societies in the future.

### 4.3.4 Stochastic ontogenetic growth model

West and West [354] generalized the OGM to incorporate the disordering influence of entropy on the growth process, by including the statistical fluctuations of the size into the dynamic equation. The dynamics of such a stochastic process eventually erases the influence of the initial state and this loss of information is a manifestation of increasing entropy. We model [354] this effect using the stochastic ontogenetic growth model (SOGM).

Uncertainty can enter into the SOGM in a number of ways, for example, through a weak coupling to the environment to produce an additive random force in the OGM as demonstrated for simple Brownian motion in classical diffusion. This is the situation we have with ontogenetic growth, where the most direct disruption is through the metabolic rate to maintain a unit biomass. We take into account the fact that this rate is not just a single number, but is a composite of multiple biological mechanisms. West et al. [369] and more recently Makarieva et al. [198] explain that the unit biomass metabolic rate incorporates, in addition to maintenance costs, the expenses associated with metabolic growth and metabolic work, etc. Consequently, rather than assuming the rate of creating a unit biomass to be a single number $\beta$, we instead assume it has a stochastic component; $\beta \to \beta_0 + \xi(t)$, where $\beta_0$ is a constant and $\xi(t)$ is a zero-centered stochastic process.

Note that the generalization of OGM made here, with a little imagination, can also be applied to UGM. However, I leave this extension of the theory to some clever graduate student in search of an interesting, but tractable, project.

Consider a phenomenological Langevin equation with multiplicative fluctuations for the TBM, where to avoid confusion we use the dummy variable $Z = M_{ij}$, which is actually the TBM for species $i$ and individual $j$. Thus, we replace equation (4.51) with the SOGM stochastic differential equation

$$\frac{dZ}{dt} = \alpha Z^b - \beta_0 Z + Z\xi(t), \tag{4.69}$$

with $\xi(t)$ given by a random process of strength $D$, with normal statistics and delta correlations given by equation (4.34). In equation (4.69), when the magnitude of $Z$ increases, the influence of the fluctuations on the growth rate are amplified and when the magnitude of $Z$ decreases, they are suppressed. The nonlinear nature of the AR makes solving this stochastic differential equation extremely difficult in general. Therefore, we turn our attention to the phase space equation of motion for the corresponding PDF $P(z,t|z_0)$ and restrict our attention to the steady-state PDF solution

$$P_{ss}(z) = \lim_{t\to\infty} P(z,t|z_0) \tag{4.70}$$

and the corresponding long-time averages of the dynamic variable

$$\langle z \rangle_{ss} = \int_{-\infty}^{\infty} z P_{ss}(z) dz. \tag{4.71}$$

An equivalent description of the SOGM is given by the corresponding evolution of the PDF in phase space. However, a Langevin equation with multiplicative fluctuations is not uniquely defined. The integral of the random force

$$\int_0^t Z(t')\xi(t')dt' \tag{4.72}$$

for a $\delta$-correlated Gaussian fluctuations, with $\xi(t)$ a non-differentiable Wiener process, can have a number of alternate definitions. As discussed by Lindenberg et al. [183], two interpretations of this integral are most frequently used: the Itô interpretation [153], where the value of the variate is independent of the Wiener interval and the Stratonovich interpretation, where the value of the variate depends on the midpoint of the Wiener interval. The difference between these two calculi can, in general, yield significantly different PDFs. In the present case, however, we are only interested in the long-time behavior of the PDF, where the two solutions coincide. So let us turn to the construction of phase space equations for the PDFs.

## 4.4 Fokker–Planck equations

In this section we examine a strategy to construct the PDF that determines the asymptotic properties of the dynamics of the SOGM. It turns out that this steady-state PDF solution to the phase space equations models the metabolic ARs as a relation between moments. To follow this modeling path requires a review of some methods from nonequilibrium statistical physics [184, 266] in an allometry context. An extension of these arguments to fractional differential equations [166, 345] is presented in Chapter 6 to make them applicable in a general allometry context.

### 4.4.1 Phase space distribution

The conditional phase space distribution function for the dynamic variable $Z_\xi(t, z_0)$, which is the solution to a Langevin equation driven by a random force $\xi(t)$, for a given initial condition $z_0 \equiv Z_\xi(0)$, can be expressed as

$$\rho_\xi(z, t|z_0) \equiv \delta(z - Z_\xi(t; z_0)). \tag{4.73}$$

The phase space distribution function is a stochastic quantity, whose average over an ensemble of realization of the random force initiated at the same starting point, yields the conditional PDF

$$P(z,t|z_0) = \langle \delta(z - Z_\xi(t;z_0)) \rangle_\xi. \tag{4.74}$$

Following Lindenberg and West [184], we construct the differential equation describing the dynamics of the phase space distribution function given by equation (4.73), and perform the average indicated by equation (4.74) to obtain the equation obeyed by the conditional PDF.

The quantity $Z_\xi(t;z_0)$ is the solution to the dynamic equations for a system open to the environment, with the SDE of the form [184, 286]

$$\frac{dZ_\xi(t)}{dt} = G(Z_\xi) + g(Z_\xi)\xi(t), \tag{4.75}$$

where $G(\cdot)$ and $g(\cdot)$ are analytic functions of their arguments. This solution can be represented by a trajectory in phase space that begins at the phase point $z_0$. Lindenberg and West show that the distribution $\rho_\xi(z,t|z_0)$ of trajectories that start at $z_0$ satisfies the continuity equation that expresses the conservation of phase space points, that is, the conservation of systems. This one-dimensional continuity equation is the Liouville equation for a compressible fluid:

$$\frac{\partial \rho_\xi(z,t|z_0)}{\partial t} + \frac{\partial}{\partial z}\left[\frac{dz}{dt}\rho_\xi(z,t|z_0)\right] = 0, \tag{4.76}$$

subject to the initial condition

$$\rho_\xi(z,t=0|z_0) = \delta(z - z_0). \tag{4.77}$$

Let $F(z)$ be an analytic function of its argument and consider the time derivative of $F(Z_\xi(t;z_0))$:

$$\frac{dF(Z_\xi)}{dt} = F'(Z_\xi)\frac{dZ_\xi(t,z_0)}{dt}, \tag{4.78}$$

where the prime denotes a derivative with respect to the function's argument. Equation (4.73) allows us to express the function as the phase space integral

$$F(Z_\xi) = \int dz \rho_\xi(z,t|z_0)F(z), \tag{4.79}$$

so that the time derivative in equation (4.78) can be expressed in terms of the phase space integral

$$\begin{aligned}\frac{dF(Z_\xi)}{dt} &= \frac{d}{dt}\int dz \rho_\xi(z,t|z_0)F(z)\\ &= \int dz \frac{\partial \rho_\xi(z,t|z_0)}{\partial t}F(z).\end{aligned} \tag{4.80}$$

The RHS of the time derivative equation (4.78) can be expressed in terms of a phase space integral in the same way

$$F'(Z_\xi)\frac{dZ_\xi(t,z_0)}{dt} = \int dz\rho_\xi(z,t|z_0)[G(z)+g(z)\xi(t)]\frac{dF(z)}{dz},$$

where we have used the Langevin equation. Integrating this last expression by parts and assuming the boundary conditions give zero contribution, yields

$$F'(Z_\xi)\frac{dZ_\xi(t,z_0)}{dt} = -\int dzF(z)\frac{\partial}{\partial z}\{\rho_\xi(z,t|z_0)[G(z)+g(z)\xi(t)]\}. \tag{4.81}$$

Comparing equations (4.80) and (4.81), as well as noting that $F(z)$ is an arbitrary function, allows us to equate the integrands and write

$$\frac{\partial\rho_\xi(z,t|z_0)}{\partial t} = -\frac{\partial}{\partial z}\{[G(z)+g(z)\xi(t)]\rho_\xi(z,t|z_0)\}, \tag{4.82}$$

which is the continuity equation with the time derivative of the variate replaced by the RHS of the multiplicative Langevin equation.

Introducing the operators $\mathcal{L}_0$ and $\mathcal{L}_\xi(t)$ that act on an arbitrary function $f(z,t)$ to yield, for the deterministic evolution of the system

$$\mathcal{L}_0 f(z,t) = -\frac{\partial}{\partial z}[G(z)f(z,t)], \tag{4.83}$$

and for the stochastic evolution of the system

$$\mathcal{L}_\xi(t)f(z,t) = -\xi(t)\frac{\partial}{\partial z}[g(z)f(z,t)]. \tag{4.84}$$

In terms of these operators the continuity equation can be written

$$\frac{\partial\rho_\xi(z,t|z_0)}{\partial t} = \mathcal{L}_0\rho_\xi(z,t|z_0) + \mathcal{L}_\xi(t)\rho_\xi(z,t|z_0) \tag{4.85}$$

whose average over an ensemble of realizations of the random force yields

$$\frac{\partial P(z,t|z_0)}{\partial t} = \mathcal{L}_0 P(z,t|z_0) + \langle L_\xi(t)\rho_\xi(z,t|z_0)\rangle_\xi \tag{4.86}$$

The form of the phase space equation for the PDF depends on the statistical properties of the random force $\xi(t)$ and the details for a Wiener process random force are given in Lindenberg and West [184] and will not be repeated here.

### 4.4.2 Solution to FPE

If we assume the fluctuations have Gaussian statistics and to be delta correlated in time, then using the Stratonovich interpretation in the reduction of equation (4.86) we obtain the Fokker–Planck equation (FPE) [352]:

$$\frac{\partial P(z,t|z_0)}{\partial t} = \frac{\partial}{\partial z}\left[-G(z) + Dg(z)\frac{\partial}{\partial z}g(z)\right]P(z,t|z_0). \tag{4.87}$$

Note that the diffusion term in equation (4.86) truncates exactly at second order for a random force modeled by a Wiener process to yield equation (4.87). However, proving this is non-trivial and we refer the interested reader to [184].

The steady-state PDF solution $P_{ss}(z)$ to the FPE can be obtained from the condition

$$\frac{\partial P_{ss}(z)}{\partial t} = 0 = \frac{\partial}{\partial z}\left[-G(z) + Dg(z)\frac{\partial}{\partial z}g(z)\right]P_{ss}(z),$$

for which the PDF is independent of both time and the initial state of the system. Imposing a zero-flux boundary condition yields

$$\left[-G(z) + Dg(z)\frac{\partial}{\partial z}g(z)\right]P_{ss}(z) = 0,$$

so that rearranging terms we obtain

$$\frac{1}{g(z)P_{ss}(z)}\frac{\partial}{\partial z}[g(z)P_{ss}(z)] = \frac{G(z)}{Dg(z)^2}. \tag{4.88}$$

Integrating this last equation yields the exact steady-state PDF [184, 286]

$$P_{ss}(z) = \frac{N}{g(z)}\exp\left[\frac{1}{D}\int\frac{G(z')dz'}{g(z')^2}\right], \tag{4.89}$$

where $N$ is the normalization.

Therefore, we choose the Stratonovich interpretation for the SOGM, with the deterministic function

$$G(z) \equiv \alpha z^b - \beta_0 z \tag{4.90}$$

and the multiplicative function

$$g(z) \equiv z, \tag{4.91}$$

where the dynamics of the PDF are described by the FPE equation (4.87), whose general solution remains a mystery. However, we can obtain the steady-state solution ($t \to \infty$), which by inserting equations (4.90) and (4.91) into equation (4.89) and carrying out the indicated operations, yields

$$P_{ss}(z) = \frac{\vartheta\gamma^{\frac{\mu-1}{\vartheta}}}{\Gamma(\frac{\mu-1}{\vartheta})}\frac{\exp[-\gamma z^{-\vartheta}]}{z^{\mu}}; \tag{4.92}$$

normalized on the interval $(0, \infty)$, with the parameter values

$$\gamma = \frac{\alpha}{\vartheta D}, \quad \mu = 1 + \frac{\beta_0}{D} \quad \text{and} \quad \vartheta = 1 - b. \tag{4.93}$$

Note that the steady-state solution to the SOGM FPE given by equation (4.92) replaces the deterministic steady-state solution obtained from the OGM [17, 369].

### 4.4.3 Fit solution to data

West and West [354] fit the parameters in equation (4.92) to a data set of mammalian species tabulated by Heusner [145]. This fit is depicted in Figure 4.4. The best data set to test this model would consist of intraspecies masses, but such a data set is not available. Therefore, we settled for second best and use an interspecies data set that was available. The data was processed by constructing a histogram of interspecies TBM for the 391 mammalian species tabulated by Heusner, by partitioning the mass axis into intervals of 20 grams and counting the number of species within each interval. The vertical axis is the relative number of species and the horizontal axis is in increments of TBM. The allometry exponent is fixed at $b = 3/4$, so that $\vartheta = 1/4$ in equation (4.92) and the remaining parameters are determined to have the mean-square, best values of $\gamma = 13.4$ and $\mu = 2.04$, with a quality of fit measured by the correlation coefficient $r^2 = 0.96$.

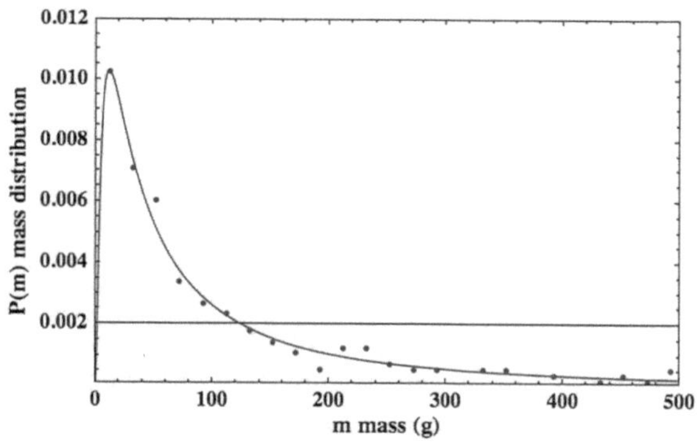


**Figure 4.4:** The histogram constructed from the average TBM data for 391 mammalian species [145] is given by the dots. The TBM data has been divided into intervals of 20 g each and the number of species within each such interval is counted. The vertical axis is the relative frequency in each interval and the solid line segment is the least squares fit of equation (4.92) to the data points. The quality of fit is $r^2 = 0.96$. (From [354] with permission.)

Figure 4.4 shows the fit to the low-mass species out to 500 grams. The remaining fit out to three million grams is included in the parameter determination, but is not shown in this figure. The asymptotic mass region is, in fact, the more interesting part of the distribution. To capture this information on a single graph we [354] construct a second histogram, this one for the asymptotic region, from approximately one thousand to three million grams. Figure 4.5 depicts the fit to the logarithmic histogram data points, indicated by the dots, starting at a TBM of 1.1 kilogram. An IPL PDF would be a straight line with a negative slope on this log–log graph. Inserting the parameter values $\mu = 1.67$ and $\gamma = 8.96$ into the steady-state TBM PDF, given by equation (4.92), yields the solid curve in Figure 4.5, which fits the data extremely well. The curve is quite clearly an IPL in the interspecies TBM. This coarse-grained description of the interspecies TBM statistics is indicative of great variability, particularly since $\mu < 2$, indicating that the variance of the interspecies TBM over an unbounded domain would diverge.

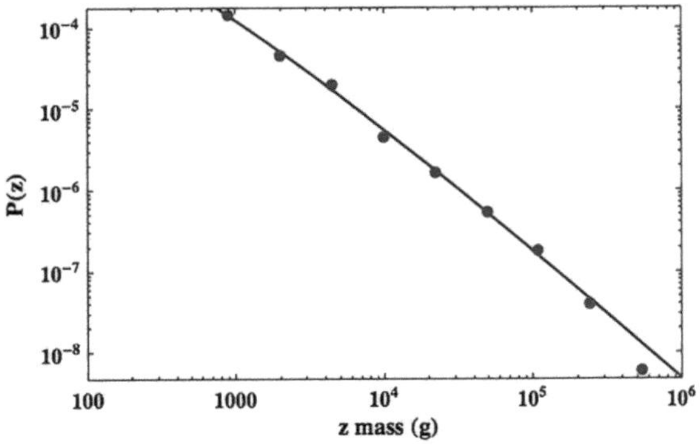


**Figure 4.5:** The average TBM data for 391 mammalian species [145] are used to construct a histogram. The mass interval is divided into twenty equally spaced intervals on a logarithm scale and the number of species within each interval is counted. The least-square fit to the nine data points is then made using logarithmically transformed distribution, see [354] for details. The quality of the fit is $r^2 = 0.998$. (From [354] with permission.)

A diverging second moment of the TBM indicates a great deal of variability in the underlying process, which in turn provides a measure of complexity of the underlying ecosystem. This PDF also provides a way to quantify the influence of one species on another, through the complexity or information gradient. Assuming that there is an optimal range of TBM for which a given species can survive, this analysis indicates that the variability of the average TBM quantifies the notion of biological fitness. This notion of fitness means the potential to survive to the age of reproduction, find a mate, and have an offspring; the greater the number of offspring, the greater the fitness.

### 4.4.4 Interspecies empirical AR

We [349, 351, 354] explicitly constructed a statistical theory of the fluctuations in the average species BMR $\langle B_i \rangle$ and average species TBM $\langle M_i \rangle$ to calculate the interspecies metabolic AR. The strategy is to relate the calculated average BMR to the average TBM through the empirical AR. The phase space approach can be followed here to obtain the long-time or steady-state average SOGM TBM, using the steady-state PDF given by equation (4.92):

$$\langle z \rangle = \int_0^\infty z P_{ss}(z) dz = \frac{\vartheta y^{\frac{\mu-1}{\vartheta}}}{\Gamma(\frac{\mu-1}{\vartheta})} \int_0^\infty z^{1-\mu} \exp\left[-y z^{-\vartheta}\right] dz$$

and replacing $z$ with the TBM notation for species $i$ and carrying out the integration, reduces the average to

$$\langle M_i \rangle = y^{1/\vartheta} \frac{\Gamma(\frac{\mu-1}{\vartheta} - \frac{1}{\vartheta})}{\Gamma(\frac{\mu-1}{\vartheta})} \approx \left[ y \frac{\vartheta}{\mu - 1} \right]^{1/\vartheta}. \tag{4.94}$$

Here we interpret the ensemble average in terms of an average over an ensemble of individual adult members of species $i$. On the RHS of this equation we have approximated the ratio of gamma functions by an asymptotic value, so that substituting the parameter values into the RHS of equation (4.94) yields

$$\langle M_i \rangle \approx \left( \frac{a}{B_m} \right)^{1/\vartheta}. \tag{4.95}$$

The equality in equation (4.95) is obtained by substituting the values of the parameters from OGM and using $b = 3/4$ and $\vartheta = 1/4$ to yield

$$B_m \approx a \langle M_i \rangle^{-1/4}. \tag{4.96}$$

Thus, the average metabolic rate necessary to maintain a unit biomass is dependent on the inverse quarter power of the average TBM of the species. This is the same expression obtained by Moses et al. [227], if we identify their adult TBM with the average TBM obtained from the steady-state PDF of the SOGM.

Note that the average TBM for species $i$ replaces $M_0$ obtained from the $dM/dt = 0$ solution to equation (4.54). On the other hand, the average BMR can be determined from the $b^{\text{th}}$ moment of the TBM

$$\begin{aligned} \langle B_i \rangle = a \langle M_i^b \rangle &= a \int_0^\infty z^b P_{ss}(z) dz \\ &= a y^{-1+1/\vartheta} \frac{\Gamma(\frac{\mu-1}{\vartheta} + 1 - \frac{1}{\vartheta})}{\Gamma(\frac{\mu-1}{\vartheta})} \approx a \left[ y \frac{\vartheta}{\mu - 1} \right]^{1/\vartheta - 1}. \end{aligned} \tag{4.97}$$

Comparing this last equation with equation (4.94) allows us to write

$$\langle B_i \rangle \approx a \langle M_i \rangle^{1-\vartheta} = a \langle M_i \rangle^b, \tag{4.98}$$

which is the interspecies metabolic AR. We emphasize [354] that the interspecies AR is a phenomenological equation that without the SOGM does not have a formal theoretical basis.

The SOGM constitutes the first theoretical construction of an interspecies metabolic AR, starting from a fundamental dynamic equation and relating the calculated averages. It is true that we used a nonlinear multiplicative Langevin equation, but the original dynamics stem from the conservation of energy argument [369] and the fluctuations are a consequence of the dynamics being degraded by entropy [351], due to interactions with the environment. This suggests a relatively untraveled path for future research in this area.

## 4.5 Summary

In this chapter we have indicated how some of the fundamental principles, applied to the development of past allometry models, have been successful in certain limited domains. The traditional optimization arguments applied to energy and efficiency have been replaced by those involving fractal geometry and the adaptation to variability. One of the most successful applications of the fractal design principle, in the physiologic domain, appears to be the WBE model, in which nature is assumed to maximize the fitness of organisms. The WBE model has the virtues of being both elegant, straightforward, and yielding the value 3/4 for the allometry exponent, independently of the branching architecture of the underlying network.

But as with most models, with this degree of generality, it is not without its critics. The first is empirical and concerns the coevolution of the allometry parameters implying that no one value of the allometry exponent is to be expected. The second, and perhaps the more compelling counter argument, is against the general use of optimization in biological models, at least when applied to deterministic quantities. The suggestion is to replace fractal geometry scaling, with that corresponding to fractal statistics, which was vindicated by the excellent fit to data for the distribution of TBM across species depicted in Figure 4.5.

The asymptotic variability of the interspecies TBM was shown to be given by the steady-state PDF solution to the FPE, resulting from the random fluctuations in the SOGM, disrupting the deterministic dynamics of the OGM. This steady-state PDF was shown to entail the interspecies empirical metabolic AR. It is worth emphasizing that this is the first theoretical derivation of the empirical AR, that is, the allometry relation involving the average BMR and average TBM. Past derivations of ARs involve the quantities themselves and not the empirically determined averages.

# 5 Strange kinetics

A dynamic fractal processes has been observed to be rich in interconnecting scales, with no scale dominating. This scale-free dynamic property is captured by the scaling behavior of the functions used to represent fractal phenomena. It has been known, since the nineteenth century mathematician, Weierstrass, answered a challenge to construct a function that was everywhere continuous, but nowhere differentiable. The challenge was made by his one time student, Cantor, of *Cantor set* fame, and Weierstrass' answer to him in 1872 was to construct the first fractal function. Since such functions have no derivatives, they cannot be solutions to traditional equations of motion, with integer-order derivatives. Consequently, fractal phenomena cannot be mechanical processes, simple or otherwise.

Thus, information in fractal phenomena is coupled across multiple scales, as we have observed in the architecture of the mammalian lung [230, 331, 338] and in the growth of cities [27]; recorded in the long-range correlations in human gait [138, 343] and the extinction record of biological species [92]; measured in the human cardiovascular network [243] and recorded in a substantial number of other contexts, as well [347]. Thus, we have both deterministic and statistical applications of the fractal concept contributing to understanding the multi-scaled phenomena that manifest allometry patterns.

The present chapter presents a strategy for the systematic treatment of dynamic fractal phenomena, using fractional differential equations (FDEs) and fractional Langevin equations (FLEs). We briefly demonstrate how to solve FDEs and explain why they appear in describing the dynamics of complex phenomena of interest in allometry. A question that immediately springs to mind is: Other than mathematicians, interested in such things out of intellectual curiosity, why would a student of science, much less a practitioner, want to learn the techniques of the fractional calculus? Or perhaps: Why has the fractional calculus languished in the mathematical backwaters for over 300 years, essentially ignored by the physical, social and life scientists? We note in passing that Leibniz was the first to implement a fractional derivative, so the fractional calculus is as old as fluxions and the differential calculus.

Direct answers to such questions are not forthcoming, but it appears that not until recently did the community of scientists see any need for a new calculus, or if occasionally seeing the need, did not openly acknowledge it. The calculus of Newton and Leibniz, along with the analytic functions entailed by solving differential equations, resulting from Newton's force laws, have historically been seen as all that is required to provide a mechanical description of the physical world. However, outside the physical sciences the mechanical portrayal of the clock-work universe has been found to fall short of a complete description of the interactions between nations, among groups that form nations, between individuals within a group, as well as, the ruminations of a person sitting alone in thought.

DOI 10.1515/9783110535136-005

We accept the physical science explanation that the myriad kinds of motion of the objects in the world around us are determined by energy gradients. Electrical energy provides the power that runs social media and lights our cities, chemical energy supplies the power to drive the engines in our transportation systems, and solar energy is converted by photosynthesis into the foods we eat. Physics provides a detailed description of how spacial changes in energy produce forces, which moves things around in space. A kite pulling at its tether, the invariant order of the colors in a rainbow, moon rise, and sunset, all have their causal explanations in terms of forces. But mechanical force laws, even when generalized to continuous media such as fluids, are not able to explain all that we observe.

We talk of individual's exerting forces on one another, as if that could be done without physical touching, or of one country exerting a force on another country, without invading. Are these non-physical forces merely metaphors, based on analogs of mechanical forces, or is there something deeper? We do not have a final answer to that question, but what we can say is that models using traditional mathematics from theoretical physics, when applied to the social and life sciences, have, with a few exceptions, been disappointing.

In this chapter, we introduce the dynamics of observables in allometry phenomena and in the next chapter we examine PDFs that can produce the AR given by equation (1.18). As stated in the previous chapter, there are two major techniques available in statistical physics for modeling dynamic complex phenomena: stochastic equations for dynamic variables, such as Langevin constructed by introducing uncertainty, through a random force, into the equations of motion; as well as, phase space equations for the PDF, using the FPE and its generalizations. The conditions under which these two strategies converge have been shown in a number of places, see, e.g., [184]. One way to apply these techniques to allometry using PDFs was discussed in the previous chapter; another, which we consider in the next chapter, requires extending both the Langevin equation and the FPE, using the fractional calculus. But before we go into these extensions it is worthwhile to consider how this 'new' calculus changes our view of the world and why it is required.

Section 5.1 examines how the scientific view of the world changes when we use the fractional calculus. We establish that the fractional derivative of a fractal function is another fractal function and therefore fractal differential equations are the necessary calculus to describe the dynamics of fractal phenomena. Section 5.2 contains a review of the solutions to simple fractional rate equations and some of the processes they describe. Perhaps more importantly, the arcane assumptions of irreversibility in time and isotropy in space are jettisoned, motivating the notion of a new kind of time, subordination time, which leads to a fractional time derivative. In Section 5.3 we discuss a generalization of a Poisson process to fractional form, using subordination time and reveal how the nature of the differential reflects the character of the process being modeled. Section 5.4 contains some speculations regarding complexity, information, the fractional calculus and how they can be interrelated.

## 5.1 Fractional thinking

For the moment let us focuses on the different ways the fractional calculus prompts us to think about complexity in all manner of phenomena and in its myriad of forms. We briefly examine the more obvious reasons why the traditional calculus, including differential equations, are insufficient to span the full range of dynamics found in natural and man-made processes and phenomena. Specifically, the complexity of nonlinear dynamic phenomena demands that we extend our horizons beyond analytic functions and analysis suggests that the functions of interest lack traditional equations of motion, such as those given by the laws of Newton. To facilitate our understanding of this lack, we introduce fractional thinking, which is a kind of in-between thinking.

Science, in one sense, has been built on the integers, such as the integer-order derivatives and integrals for deterministic processes and integer-order moments, such as the mean and variance, using the LFE. However, between these integer moments there are fractional moments, required when empirical integer moments fail to converge, that is, where the LFE breaks down. The mathematical existence of non-integer moments have always been known to exist, but there was no clear necessity to use them. However, the divergence of low-order moments has been found to be quite common for IPL PDF, which characterize the statistical variability of complex phenomena. Consequently, fractional moments have now become important.

Integer-valued operators are local and isotropic in space and time and these were the operators that dominated science and engineering for centuries. In the *Principia*, as mentioned earlier, Newton made the explicit assumptions that time was absolute and that it proceeds in a featureless flow, without disruption. In a similar way he considered space to be everywhere the same and that it only provided a context in which to put objects. Consequently, absolute space and time were of no physical significance to Newton, or the following generations of scientists, until Einstein turned the world of physics up-side-down, at the turn of the twentieth century. However we are not interested in the extremes of space and time, the very large (or small), or the very fast, and restrict our attention to those things that influence our everyday lives. In the world of the ordinary, where cars get stuck in mud, peanut butter clings to the roof of your mouth, and you remember being embarrassed in high school; in this world, Newton's laws require re-examination. Non-integer operators are necessary to describe dynamics, which have long-time memory and spatial heterogeneity. Complex phenomena require new ways of thinking quantitatively and the fractional calculus provides one framework for that thinking [359, 362].

One way to force a change in how we think is to follow a little white rabbit down a hole, as did Alice in the nineteenth century and which we describe in the *Fractional Calculus View of Complexity, Tomorrow's Science* [362]. Recall that Alice's adventures in *Wonderland* were about how different the rules of logic, on which her society was based, were from those in Wonderland. How a little English girl coped

**Figure 5.1:** In this classic view of the tea party, Alice sees a simple cube when the Mad Hatter, March Hare and Door Mouse, discuss a fractal dimensional Seripinski gasket. Such different views of the world are common place. (From [362] with permission.)

with such changes are depicted in the woodcut in Figure 5.1. On the other side of the looking glass, she was at first confused by the apparent lack of rules governing the world in which she found herself. Things that were relatively simple back home, seemed unnecessarily complicated in Wonderland. However, after some time, she understood that rules did exist; they were just very different from those that determined the world she had left behind. Much like Wonderland it is not that the quantitative reasoning discussed here does not have rules; it is just that the rules of quantification are very different from those we were taught ought to determine how the world works. The present chapter is about how the rules have changed; what that change implies about the phenomena they describe and how we are to understand them. In short, my purpose is to convince you that we have been living in Wonderland all along.

Complexity has been repeatedly emphasized in earlier chapters, along with the not so subtle implication that this is the reason for the need to have a fractional calculus. This observation provokes the question: Why is the fractional calculus entailed by complexity? The easy answer is that it is not. The fractional calculus is not required to understand any specific instance of complexity, in fact, most investigators have relied on traditional calculus to provide their explanations of complex phenomena. However, as explained in our book on the fractional calculus [362]:

> ...we interpret this lack of uniqueness in the connection between complexity and the fractional calculus in the same way we do that of Newton not explicitly using, his then newly formulated fluxions, in his discussion of mechanical motion in the *Principia*. Rather than using fluxions, his version of differentials, he couched his arguments in geometry, the mathematical/scientific language of the day. It must be noted that the underlying fluxion-based infinitesimals were disguised as traditional, often impenetrable, geometrical arguments that relied heavily on geometrical limit-increments of variable line segments. In an analogous way, I believe that many of the complex phenomena that have required often tortuous explanations, using traditional methods, will eventually be shown to be more naturally described using the fractional calculus. This is a statement of a research vision that has only been partially realized [195, 251, 345].

We left complexity vaguely defined in Section 1.3 and will continue to leave it so. Instead of providing an unnecessarily restrictive definition of complexity, we suggest a working definition by considering phenomena, or structures, to be complex, when traditional analytic functions are not able to capture their richness in space and/or time. It is traditional wisdom that a physical theory, such as classical mechanics, can be used to predict, with absolute certainty, the dynamics of highly idealized systems, using limiting concepts abstracted from the real world. However, that bubble, regarding the sacrosanct character of physical theory, was burst in 1888 by Poincaré [252]. It took nearly a century for the significance of Poincaré's revolutionary mathematics to penetrate the insulation of mainstream physics. None-the-less the nonlinear terms in a Hamiltonian were eventually shown to produce non-predictability and to lead to non-integrable Hamiltonian systems in which particle trajectories breakup into a random spray of points in phase space. This is certainly one kind of complexity. Another is the kind of chaos that arises in dissipative dynamical systems, where the dynamics unfold on strange attractors, that is, an attractor with a fractal dimension, see, e.g., [236].

### 5.1.1 Dynamic fractals

The fractional calculus was not completely ignored by science and engineering, but it must be stressed that it only appeared in studying phenomena where all other options had been exhausted, as in determining the properties of viscoelastic materials; materials whose dynamics are neither the same as solids nor fluids, but lie somewhere between the two. Consequently, the first scientific application of the fractional calculus facilitated the understanding of the dynamics of materials such as tar, hemoglobin, polymers and taffy, to name a few. The equations of motion for such materials fall outside traditional Newtonian dynamics and fluid mechanics, because the material properties are neither those of solids nor fluids [260]. Historically the viscoelastic equations of motion were shown to be of an integro-differential form and were subsequently revealed to have an equivalent interpretation in terms of fractional dynamics [120, 292]. To understand the reason for this equivalence, we examine the notion of dimension more carefully.

Euclid organized our understanding of physical structures into classical geometry over two millennia ago. In so doing he gave us the metrics of points, lines, planes and other, now familiar, geometrical objects. Mandelbrot [201], during my lifetime, has pointed out the obvious: lightening does not move in straight lines, clouds are not spheres and most physical phenomena violate the underlying assumptions of Euclidean geometry. Not given to half-measures, however, Mandelbrot relentlessly pursued his ideas to their logical conclusions. He introduced the idea of a fractal dimension into the scientific lexicon and proceeded to catalogue the myriad of naturally occurring complex phenomena that ought to be described by his fractal geometry and subsequently by fractal statistics. We discussed some of these things briefly in the first chapter.

The application of fractals exploded into a vast array of disconnected fields of study in the late 1970s and through the 1990s, from statistical physics [214], to the branching network of the mammalian lung [331, 338], to the growth patterns of urban centers [27], and to intermittent search strategies for prey and lost individuals [33]. In these and many other studies it became apparent that not just static structures are fractal, but the dynamics of complex phenomena are fractal as well, including their statistical fluctuations [349, 359]. It can be argued that the development of the fractal concept and its application to dynamic phenomena paved the way for the acceptance of fractional equations of motion, in particular, and the fractional calculus, in general.

Consider two points on a plane separated by a finite Euclidean distance. These two points can be connected by a line segment of finite length, one that becomes longer and longer as the variability of the line increases. Fractal geometry maintains that a continuous line, connecting the two points, can be constructed that is infinitely long. The existence of such a line can be demonstrated using a prototypical fractal function constructed by Weierstrass as a Fourier series. His function has the property of being continuous everywhere, but being nowhere differentiable. A century later Mandelbrot [201] generalized the Weierstrass function (GWF) to the form:

$$W(t) \equiv \sum_{n=-\infty}^{\infty} \frac{1}{\alpha^n}[1 - \cos(\beta^n \omega_0 t)] \tag{5.1}$$

with the parameter values $\beta > \alpha > 1$. Note that the degree of variability is determined by the powers of the parameter $\beta$ and the corresponding amplitude of that variability is determined by the powers of the parameter $\alpha$. The relative size of these two parameters yields the fractal dimension of the curve, as shown below. Note the similarity to the renormalization group transformation argument used in the WBG model presented in Chapter 4. An example of the GWF is graphed in Figure 5.2 for specific values of $\alpha$ and $\beta$.

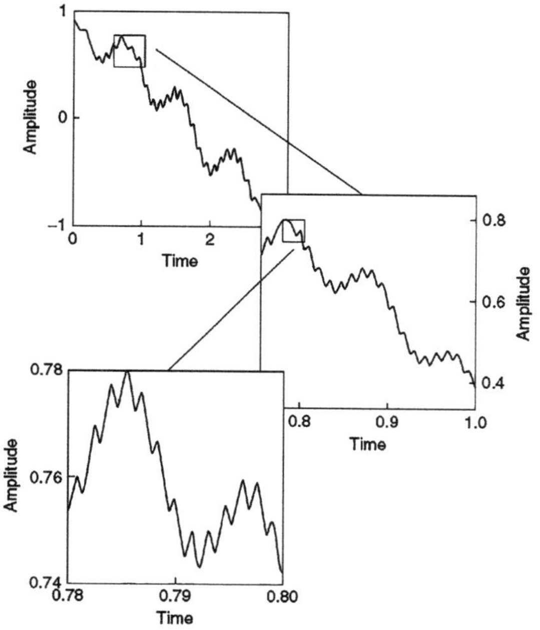


**Figure 5.2:** The GWF defined by equation (5.1) is graphed for the parameter values $a = 4$ and $b = 8$. Note the magnified region of a section of the top curve, reproduces that curve, but on a different scale. The same can be said for the magnified region of the second curve. The degree of magnification can be traced by examining the units of each rendition.

Such a fractal curve has the property of being self-similar on all scales, that is, in the neighborhood of any point along the curve there is variability. In fact the curve has such a richness of scales that under increasing magnification, in the vicinity of that point, there is variability, no matter how close to the point the curve is observed. Consequently, it is not possible to draw a tangent to the curve at any point along the curve and therefore the continuous function is neither differentiable, nor analytic. As the limit is taken in the vicinity of any point, more and more structure of the curve reveals itself, and the derivative of the curve becomes ill-defined. This property was anticipated by Perrin [245] in his experimental study of diffusion, see Feder [99] and Mandelbrot [201] for a complete inventory of fractal properties. Subsequently, it was argued that the equations of motion for such complex phenomena must be fractional [133, 274, 345].

The fractal properties of the GWF, including its fractal dimension, are determined using a scaling argument. The scaling is determined by multiplying the independent variable $t$ by the parameter $\beta$:

$$W(\beta t) \equiv \sum_{n=-\infty}^{\infty} \frac{1}{\alpha^n}[1 - \cos(\beta^{n+1}\omega_0 t)] = \alpha W(t). \tag{5.2}$$

The second equality is obtained by re-indexing terms in the infinite series, resulting in a RG equation. The solution to the RG equation is of the scaling form [214, 352], as discussed in Chapter 1. The scaling in equation (5.2) is the analytic manifestation of the self-similarity property of the GWF. It is clear from Figure 5.2 that if any segment of the curve is magnified, the entire curve is again revealed, due to self-similarity. Finally the time derivative of the GWF can be shown to diverge for $\beta > \alpha$ [201] Due to the divergence of the integer time derivative, this function cannot be the solution to any traditional equation of motion. The Weierstrass function was considered to be a mathematical curiosity and prior to Mandelbrot's analysis, was thought not to describe any 'real' process, with one notable exception, which we now consider.

The potential physical significance of the Weierstrass function was first recognized by Richardson [268], who had measured the increasing span of plumes of smoke ejected from chimneys. The dispersion of the smoke plume was driven by fluctuating wind fields over the tops of buildings, see Figure 5.3. From his observations he speculated that the Weierstrass function might characterize the turbulent wind speed, which was known to be non-differentiable. This conjecture was motivated, in part, by the observation that the down-wind span of the plume increased in time as $t^{\mu}$ with $\mu \geq 3$, a value inconsistent with molecular diffusion for which $\mu = 1$. Half a century later Mandelbrot established that turbulent velocity fields are fractal statistical processes and that the eddy cascade model of turbulence, invented by Kolmogorov [170], was, in fact, a dynamic fractal, so that turbulence has no characteristic scale in either space or time. These lack of scales argued against the traditional equation of fluid dynamics to describe turbulent fluid flow [296].

**Figure 5.3:** The smoke plume from a chimney is seen to expand transversely with distance from the source. In 1926 Richardson [268] speculated that this expansion could be described by a Weierstrass function, since the turbulent velocity field of the wind was empirically known to be non-differentiable.

The story of the fractal description of turbulence has been repeated in multiple disciplines, over the past forty years; not the details, but the generic idea that complex phenomenon have self-similar dynamics and therefore do not possess characteristic scales in either space or time. Identifying self-similarity as a part of the process, in turn, has led to the reinterpretation of existing data sets using fractal geometry, fractal dynamics and/or fractal statistics. Scaling has been crucial and appears in multiple forms; reaching back over five hundred years to da Vinci's branching relation [269]; fast forwarding four hundred years to Pareto's IPL distribution of income [240]; up to the contemporary PDFs for the occurrence of wars of a given size [273] and the scale-free character of complex networks [4].

The oscillatory coefficient in the scaling solution of the GWF given in equation (1.5) has been shown to be a consequence of the underlying process having a complex dimension [297, 309]: $D = D_R + iD_I$, where both the real and imaginary dimensions are real. The parameter $D_R$ is related to the index of the IPL of the phenomenon being measured and $D_I$ is a measure of the period of logarithmic modulation of the IPL. Such complex dimensions have been observed in the architecture of the human lung [297, 338], as well as, in earthquakes, turbulence, and financial crashes [352, 362]. The most important point for us here is that once a scale-free process has been identified as being described by a fractal function, we know that the traditional calculus will not be available to determine the dynamics of the process the fractal function represents, because the integer-derivatives of fractal functions diverge. This is where fractional operators enter our thinking.

### 5.1.2 Simple fractional operators

We have written an essay motivating the use of the fractional calculus [362] to understand complex phenomena and what you are reading is not it, although there is some unavoidable, but modest, overlap. Here we are content to lay out some of the more exotic features of this unfamiliar calculus, along side their most useful applications for allometry science. Starting from the properties of Gamma functions extended into the complex plane, it is possible to write the derivative of a monomial as follows:

$$\frac{d^\alpha}{dt^\alpha}[t^\beta] = \frac{\Gamma(\beta+1)}{\Gamma(\beta+1-\alpha)} t^{\beta-\alpha}. \tag{5.3}$$

This is not a particularly surprising equation, since if $\alpha$ and $\beta$ are integers the expression gives nothing more than the sequential derivative for $\beta > \alpha$

$$\frac{d^\alpha}{dt^\alpha}[t^\beta] = \beta(\beta-1)\cdots(\beta-\alpha+1)t^{\beta-\alpha}.$$

However, if we stipulate that the order of the derivative $\alpha$ is not an integer, something remarkable happens.

Consider the case of $\alpha = 1/2$ for various values of $\beta$:

$$\frac{d^{1/2}}{dt^{1/2}}[t^{-1/2}] = 0, \quad \beta = -1/2; \tag{5.4}$$

$$\frac{d^{1/2}}{dt^{1/2}}[1] = \frac{1}{\sqrt{\pi t}}, \quad \beta = 0; \tag{5.5}$$

$$\frac{d^{1/2}}{dt^{1/2}}[t] = \sqrt{\frac{t}{\pi}}, \quad \beta = 1. \tag{5.6}$$

The result given by equation (5.4) is strange in that it is at odds with what we have been taught for integer derivatives. Every first year student of the calculus knows that only the derivative of a constant is zero for arbitrary values of the variable. Furthermore, the finite derivative of a monomial cannot be zero for non-zero values of the variable, unless $\alpha > \beta > 0$. The fact that a derivative of a monomial vanishes is a consequence of the divergence of the Gamma function in the denominator in equation (5.3) and seems to be wrong, but it is not.

On the other hand, equation (5.5) is even stranger, given that the fractional time derivative of a constant appears to yield a time-dependent function; a result that certainly violates everything we have ever learned about differentials. Of course, neither of these curious findings is consistent with the ordinary calculus and as we subsequently show, they have to do with the non-local nature of fractional derivatives. A property at odds with the ordinary calculus, which is local by construction.

Finally, the expression given by equation (5.6) was first obtained by Leibniz, in answer to a question posed by de l'Hôpital in a 1695 letter. This is the least surprising of the three results, in that the exponent of the monomial, after taking the derivative, is lowered from its original value by the order of the derivative. But then there is the constant coefficient $1/\sqrt{\pi}$.

The three fractional derivatives have been chosen to emphasize that we are now squarely in Wonderland. A land where the rules for quantitative analysis are distinctly different from what they were back home, but they are not altogether arbitrary. What remains to be seen is whether this, in fact, is the world in which we have lived all along. But most of all, we intend to show that the fractional calculus can describe the complexity we see all around us in the physical, biological and social worlds, and which is manifest in the relations of allometry science.

We emphasize that there is no single fractional calculus, just as there is no single geometry. Different definitions of fractional differential and integral operators, have been constructed to satisfy various constraints. There are a number of excellent texts that review the mathematics of the fractional calculi [218, 280], others that emphasize the engineering applications of the fractional operators [195, 251], and still others that seek physical interpretations of the fractal operators [345, 359, 362]. In other words the literature is much too vast to cover here even in a superficial way. However, many insights into complexity can be made by judiciously choosing various forms of the fractional operators to discuss, which have been used in a variety of specific applications.

We denote the Laplace transform of a time-dependent function $X(t)$ by

$$\widehat{X}(u) = \mathcal{LT}\{X(t);u\} \equiv \int_0^\infty e^{-ut}X(t)dt, \tag{5.7}$$

where $u$ is the Laplace variable. The Laplace transform of the Caputo fractional derivative of order $\gamma$ [60] of the function $X(t)$ for $0 < \gamma < 1$ is,

$$\mathcal{LT}[\partial_t^\gamma[X(t)];u] \equiv u^\gamma \widehat{X}(u) - u^{\gamma-1}X(0). \tag{5.8}$$

The explicit time-dependent form of the Caputo fractional derivative is obtained by taking the inverse Laplace transform of equation (5.8):

$$\partial_t^\gamma[X(t)] = \frac{1}{\Gamma(\gamma)}\int_0^t \frac{dt'}{(t-t')^\gamma}\frac{dX(t')}{dt'}. \tag{5.9}$$

Note that the function must have a first-order derivative with respect to the argument, which is not required for other definitions of the fractional derivative.

It is now possible to return to the GWF and examine its Caputo fractional derivative using equation (5.8). Thus, we obtain

$$\mathcal{LT}[\partial_t^\gamma[W(t)];u] \equiv u^\gamma \widehat{W}(u), \tag{5.10}$$

since $W(0) = 0$. The operation of taking the fractional derivative of the GWF can be determined by taking the inverse Laplace transform of equation (5.10) to define:

$$\begin{aligned} W^\gamma(t) &= \mathcal{LT}^{-1}[u^\gamma \widehat{W}(u);t] \\ &= \mathcal{LT}^{-1}\left[\sum_{n=-\infty}^{\infty}\frac{(\beta^n\omega)^2}{\alpha^n}\frac{u^{\gamma-1}}{u^2+(\beta^n\omega)^2};t\right] \end{aligned} \tag{5.11}$$

The fractional derivative of the GWF can then be shown, by rearranging terms in the series, to satisfy the RG relation:

$$W^\gamma(\beta t) = \frac{\alpha}{\beta^\gamma}W^\gamma(t), \tag{5.12}$$

which is solved, just as we did previously, to obtain

$$W^\gamma(t) = A(t)t^\mu, \quad \mu = \frac{\log\alpha}{\log\beta} - \gamma, \tag{5.13}$$

where again the function $A(t)$ is Fourier series in the variable $\log t$. Comparing the scaling index in equation (1.5), with that in equation (5.13), we see that taking the fractional derivative shifts the scaling parameter by a factor $\gamma$ and the resulting function can be shown to converge [274]. Thus, the fractional derivative of the GWF, unlike the integer derivative of this fractal function, does not diverge.

A fractional operator of order $\gamma$, acting on a fractal function of fractal dimension $D$, in the case of the GWF the fractal dimension is expressed in terms of the scaling parameters, $D = \log\alpha/\log\beta$, consequently yields another fractal function with fractal dimension $D \pm |\gamma|$. The fractional operator can be either a derivative or an integral; the former increases the fractal dimension, thereby making the function more erratic; whereas the latter decreases the fractal dimension, thereby making the function smoother. The fact that the fractional operator acting on a fractal function converges, supports the conjecture that the fractional calculus can provide an appropriate description of the dynamics for fractal phenomena [133, 274].

We can strengthen this argument regarding the dynamics of fractal processes by recalling the definition of a derivative to an analytic function represented by a Taylor series

$$F(t) = F(\tau) + \sum_{j=1}^{\infty} C_j (t-\tau)^j. \tag{5.14}$$

The first derivative of the Taylor series at $t = \tau$ determines the lowest-order coefficient in the series

$$\begin{aligned} F^{(1)}(\tau) &= \lim_{\Delta t \to 0} \frac{F(t+\Delta t) - F(t)}{\Delta t}\bigg|_{t=\tau} = \sum_{j=1}^{\infty} C_j \lim_{\Delta t \to 0} \frac{(t-\tau+\Delta t)^j - (t-\tau)^j}{\Delta t}\bigg|_{t=\tau} \\ &= C_1. \end{aligned}$$

In a similar way the higher-order coefficients in the series are obtained by the higher-order derivatives evaluated at $t = \tau$ to obtain

$$F^{(n)}(\tau) = \Gamma(n+1) C_n$$

and the Taylor series is expressed explicitly in terms of integer-order derivatives

$$F(t) = F(\tau) + \sum_{j=1}^{\infty} \frac{F^{(n)}(\tau)}{\Gamma(n+1)} (t-\tau)^j. \tag{5.15}$$

Such analytic functions determine the behavior of a great many physical phenomena and their development was a focal point of science in the nineteenth century.

However, a number of physical phenomena, such as turbulence and phase transitions, cannot be represented using analytic functions. So let us examine a more general kind of function, say one defined by a fractional Taylor series:

$$G(t) = G(\tau) + \sum_{j=1}^{\infty} C_j (t-\tau)^{j\alpha}, \tag{5.16}$$

with $0 < \alpha < 1$. The first-order derivative of this function again defines the lowest-order coefficient in the series

$$G^{(1)}(\tau) = \lim_{\Delta t \to 0} \left. \frac{G(t+\Delta t) - G(t)}{\Delta t} \right|_{t=\tau} = \sum_{j=1}^{\infty} C_j \lim_{\Delta t \to 0} \left. \frac{(t-\tau+\Delta t)^{j\alpha} - (t-\tau)^{j\alpha}}{\Delta t} \right|_{t=\tau}$$
$$= \sum_{j=1}^{\infty} C_j \lim_{\Delta t \to 0} \frac{\Delta t^{\alpha} + \cdots}{\Delta t} = \infty,$$

where the terms in the expansion are not indicated, since they are higher-order than $\Delta t$. The lowest-order term $\Delta t^{\alpha-1}$ clearly diverges in the limit for $C_1 \neq 0$. It is evident that such a function cannot be the solution to the traditional, integer-order equation of motion.

On the other hand, we can define the $\beta$-order derivative as

$$G^{(\beta)}(\tau) = \lim_{\Delta t \to 0} \left. \frac{G(t+\Delta t) - G(t)}{\Delta t^{\beta}} \right|_{t=\tau} = \sum_{j=1}^{\infty} C_j \lim_{\Delta t \to 0} \left. \frac{(t-\tau+\Delta t)^{j\alpha} - (t-\tau)^{j\alpha}}{\Delta t^{\beta}} \right|_{t=\tau}$$
$$= \sum_{j=1}^{\infty} C_j \lim_{\Delta t \to 0} \frac{\Delta t^{\alpha} + \cdots}{\Delta t^{\beta}} = C_1 \delta_{\alpha,\beta}$$

to determine the series coefficients. Consequently, the fractional derivative of the fractional Taylor series is finite and the series coefficients can be expressed in terms of the fractional derivatives at the point of expansion of the series

$$G(t) = G(\tau) + \sum_{j=1}^{\infty} \frac{G^{(j\alpha)}(\tau)}{\Gamma(j\alpha+1)} (t-\tau)^{j\alpha}. \quad (5.17)$$

The convergence of the fractional derivatives to finite values for the coefficients suggests that the fractional calculus can be used to define the fractional equations of motion to describe the dynamic behavior of a fractal process. We shall use the fractal Taylor series to solve simple fractional rate equations in the next section.

## 5.2 Fractional rate equations

Let us now consider the difference between an ordinary rate equation and a fractional rate equation. A simple dynamic process is described by the relaxation rate equation for the dynamic variable $Q(t)$:

$$\frac{dQ(t)}{dt} = -\lambda Q(t), \quad (5.18)$$

where $\lambda$ is the rate of relaxation for the dynamic variable $Q(t)$. The solution to this equation is, of course, given by the exponential relaxation from the initial condition $Q(0)$ to zero

$$Q(t) = Q(0)e^{-\lambda t}. \quad (5.19)$$

This is the unique solution to this initial value problem and provides everything we can know about the system.

We can also interpret $Q(t)$ as the probability of the occurrence of an event, such as the decay of a radioactive particle, by interpreting the initial condition as $Q(0) = 1$, so the solution is the probability of a decay occurring at time $t$, given that a decay occurred at time $t = 0$. It can describe the probability of failure of a widget coming off an assembly line, or the relaxation of a perturbation to a linear system. It is also the simplest dynamic response of a system to a weak stimulus. Consequently, the rate equation describes the generation of decay events by a Poisson process, as we discuss subsequently.

Of course, the relaxation of disturbances in complex materials, such as taffy or rubber are not described by equation (5.18) and were discovered to require a fractional relaxation equation to describe the decay to a steady state. One derivation of the fractional calculus representation of relaxation is based on the notion of self-similar dynamics, as manifest through renormalization behavior. Glöckle and Nonnenmacher [120] argue that the RG concept may be applied to the rate equation by assuming the existence of many conformational sub-states separated by energy barriers. They assume a dichotomous stochastic process in which the relaxation between two states is not given by a single rate $\lambda$, but by a distribution of rates such that the relaxation of a function is given by

$$Q(t) = \int_0^\infty \rho(\lambda) \exp[-\lambda t] d\lambda, \tag{5.20}$$

where $\rho(\lambda)$ is the distribution of rates that represent the reaction kinetics and relaxation. However, there are many other phenomena that can be modeled in this way, including thermally activated escape processes [69], intermittent fluorescence of single molecules [136] and nanocrystals [46], stochastic resonance [107] and blinking quantum dots [161], to name but a few. We subsequently show two other ways of arriving at a fractional differential equation.

Glöckle and Nonnenmacher [121] introduced a fractal scaling model for the distribution of reaction rates, from which they were able to derive the fractional differential rate equation:

$$D_t^\alpha[Q(t)] - \frac{t^{-\alpha}}{\Gamma(1-\alpha)} Q(0) = -\lambda_0 Q(t), \tag{5.21}$$

where $\lambda_0$ is now the constant relaxation rate. Here the Riemann–Liouville (RL) fractional operators are introduced. The RL integral is defined:

$$D_t^{-\alpha}[Q(t)] \equiv \frac{1}{\Gamma(\alpha)} \int_0^t \frac{Q(t')dt'}{(t-t')^{1-\alpha}}, \tag{5.22}$$

from which the RL fractional derivative is defined:

$$D_t^\alpha[Q(t)] \equiv D_t^n D_t^{\alpha-n}[Q(t)], \tag{5.23}$$

and the operator index is in the range $n-1 \leq \alpha \leq n$ for integer $n$. Note that the fractional relaxation equation (5.21) reduces to the ordinary one equation (5.18) when $\alpha = 1$.

A simple exercise is to use the RL fractional derivative to derive the derivatives in equation (5.3). Try it and lose some of the fear associated with manipulating integral operators.

An alteration of the RL fractional derivative, which can be used to describe the dynamics of a non-differentiable process, was introduced by Jumarie, who argued that the reason for introducing fractional derivatives is to deal with non-differentiable functions [160]. He suggested the RL fractional derivative be modified to include the initial value:

$$\mathcal{D}_t^{\alpha}[Q(t)] \equiv \frac{1}{\Gamma(1-\alpha)} \frac{d}{dt} \int_0^t \frac{dt'}{(t-t')^{\alpha}} [Q(t') - Q(0)], \quad 0 < \alpha \leq 1, \tag{5.24}$$

which has the virtues of both the Caputo and RL fractional derivatives, without the counter-intuitive limitations. The modified RL (MRL) fractional derivative is defined for arbitrary non-differentiable functions (there are no requirements on the derivatives of the function) and when applied to a constant yields zero. Moreover, the Laplace transform of the MRL derivative of a function coincide with that of the Caputo fractional derivative.

It is reasonable to demonstrate that the MRL fractional derivative enables us to explicitly solve the fractional rate equation (FRE). To do this we follow Jumarie [160] and consider the FRE for $0 < \alpha < 1$:

$$\mathcal{D}_t^{\alpha}[Q(t)] = -\lambda_0 Q(t), \tag{5.25}$$

which using the definition of the MRL fractional derivative is

$$\frac{1}{\Gamma(1-\alpha)} \frac{d}{dt} \int_0^t \frac{dt'}{(t-t')^{\alpha}} [Q(t') - Q(0)] = -\lambda_0 Q(t). \tag{5.26}$$

Introducing the new variable $z = t'/t$ into the equation and integrating over time allows us, after simplification, to write the FRE in the form

$$\int_0^1 \frac{dz}{(1-z)^{\alpha}} [Q(zt) - Q(0)] = -\lambda_0 \Gamma(1-\alpha) t^{\alpha} \int_0^1 dz Q(zt). \tag{5.27}$$

We assume a solution to equation (5.27) of the form of a fractional Taylor series

$$Q(t) = \sum_{k=0}^{\infty} Q_k t^{k\alpha} \tag{5.28}$$

as suggested by the factor $t^{\alpha}$ multiplying the integral on the RHS of equation (5.27). Inserting this series into equation (5.27), integrating and equating coefficients of equal powers of the time yields,

$$Q_{k+1} = -\lambda_0 \frac{\Gamma(k\alpha + 1)}{\Gamma(k\alpha + \alpha + 1)} Q_k,$$

which when iterated to the initial value $k = 0$ provides the expansion coefficients in terms of the initial value $Q_0 \equiv Q(0)$

$$Q_k = \frac{(-\lambda_0)^k}{\Gamma(k\alpha + 1)} Q_0.$$

Inserting this last expression into equation (5.28) yields

$$Q(t) = Q_0 \sum_{k=0}^{\infty} \frac{(-\lambda_0 t^\alpha)^k}{\Gamma(k\alpha + 1)}, \tag{5.29}$$

the series expression for the Mittag-Leffler function (MLF).

The initial value solution to the FRE, either equation (5.21) or (5.25), was obtained by the mathematician Mittag-Leffler [220] at the opening of the twentieth century:

$$Q(t) = Q_0 E_\alpha(-\lambda_0 t^\alpha) \tag{5.30}$$

in terms of the infinite series that now bears his name

$$E_\alpha(z) = \sum_{k=0}^{\infty} \frac{z^k}{\Gamma(k\alpha + 1)}. \tag{5.31}$$

It is clear that the exponential simplicity of radioactive decay, or a failure mode of a new product, is here replaced by a more complex decay process, but the exponential simplicity is regained when $\alpha = 1$.

The time dependence of the Mittag-Leffler function (MLF) is depicted in Figure 5.4. At early times the MLF has the analytic form of a stretched exponential

$$\lim_{t \to 0} E_\alpha(-\lambda_0 t^\alpha) = \exp\left[-\frac{\lambda_0}{\Gamma(\alpha + 1)} t^\alpha\right], \tag{5.32}$$

as indicated by the long-dashed line segment in the figure, which deviates from the exact MLF solution at long times. In rheology the dashed curve is the Kohlrausch–Williams–Watts law for stress relaxation that has been fit to experimental data for over a century. Asymptotically in time the MLF yields an IPL

$$\lim_{t \to \infty} E_\alpha(-\lambda_0 t^\alpha) \propto \frac{1}{t^\alpha}, \tag{5.33}$$

as shown by the dotted line segment in Figure 5.4, which deviates from the exact MLF solution at early times. The dotted line corresponds to the Nutting law of stress relaxation. The relation of the fractional relaxation equation and its solution, the MLF, to these two empirical laws for stress relaxation was explored by Glöckle and Nonnenmacher [121]. Note that the MLF smoothly joins these two empirical regions with a single analytic function. So analytic functions are still useful they just do not appear as solutions to familiar differential equations.

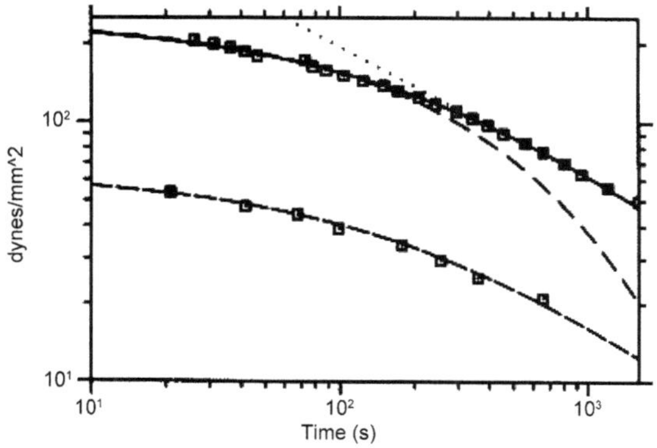


**Figure 5.4:** Stress relaxation at constant strain for two different initial conditions are indicated by the discrete data points. Upper: The solid curve is the MLF. The dotted curve is the IPL and the long-dashed curve is the stretched exponential. Lower: The dashed curve is the MLF and the boxes are data. The two fits use the MLF with different parameter values. (From [120] with permission.)

Of course, this fractional generalization of the stress relaxation equations is phenomenological rather than fundamental. The empirical law that stress is proportional to strain for solids was provided by Hooke. For fluids Newton proposed that stress is proportional to the first derivative of strain. Scott Blair et al. [292] suggested that a material, with properties intermediate to that of a solid and a fluid, for example, a polymer, should be modeled by a fractional derivative of the strain, leading to a fractional equation of the form given by equation (5.25).

The conjecture of Scott Blair was vindicated by the experimental results depicted in Figure 5.4, where data from stress relaxation experiments using polyisobutylene are shown to be well fit by the MLF. Glöckle and Nonnenmacher [121] have also compared the theoretical results with experimental data sets obtained by stress-strain experiments carried out on polyisobutylene and natural rubber and have found agreement over more than 10 orders of magnitude. They have also successfully modeled self-similar protein dynamics in Myoglobin [121] and to formulate slow diffusion processes in biological tissue [172].

The asymptotic form of the Mittag-Leffler function is an IPL as noted above. This asymptotic behavior suggests that perhaps many of the data sets that have been modeled strictly in terms of IPLs may in fact be more accurately modeled using an MLF when examined more carefully. This, in itself, suggests a direct connection between allometry science and the fractional calculus.

### 5.2.1 Another source of fractional derivatives

A general aspect of complexity is revealed through cooperative behavior emerging in complex networks. The flocking of birds [63], the schooling of fish [164] the swarming of insects [385]; the epidemic spreading of diseases [23]; the spatiotemporal activity of the brain [30, 68, 102]; the flow of highway traffic [15]; and the cascades of load shedding on power grids [61]; all demonstrate collective behavior reminiscent of particle dynamics near a critical point, where a dynamic system undergoes a phase transition. Thus, the macroscopic fluctuations observed in complex networks display emergent properties of spatial and/or temporal scale-invariance, manifest in IPLs of connectivity, as well as, in waiting-time PDFs. These IPLs cannot be inferred from the equations describing the nonlinear dynamics of the individual elements of the network. Despite the advances made by RG analyses and self-organized criticality theories, which have shown how scale-free phenomena emerge at critical points, the issue of determining how the emergent properties influence the micro-dynamics at criticality is only partly resolved [359]. However, one thing seems clear. Time, as measured by the solitary individual, is distinct from the time measured by the collective; the source of that difference lies in the influence the collective has on the perception of time by the individual.

Svenkeson et al. [311] point out that operational time, which is the time experienced by an isolated individual, determines that the individual's dynamics is deterministic. But to an experimenter, observing the individual, that same temporal behavior can appear to grow erratically, then abruptly to freeze in different states for extended time intervals, only to be released and grow again. This notion of subjective time is a concept with a long lineage, dating back to the mid-nineteenth century psycho-physical experiments and theory of Weber and those of Fechner.

Here we introduce the notion of a complex dynamic network, consisting of a large number of nonlinearly interacting individuals. Suppose that each person in isolation vacillates between two states, according to a simple rate equation. When the individuals are isolated, that is, non-interacting, the operational and chronological times coincide with Newton's mathematical time. The average behavior of the individual in this uncoupled case is specified to be given by a Poisson distribution, that is, the number of times the individual switches between the two states has a Poisson PDF. When the individuals in the network are allowed to interact, the behavior of the individual changes and the two kinds of time become distinct. The dynamics of the complex network influences the simpler dynamics and statistics of the individual. Due to the random nature of the evolution of the individual in chronological time, the subordination process involves an ensemble average over many realizations of the individual, each evolving according to her own internal clock. The resulting ensemble average over a large number of distinct realizations results in an average over a collection of independent random trajectories.

In the operational time frame, the typical temporal behavior of a person is regular and evolves exactly according to the ticks of that person's internal clock. Therefore, it is assumed that an individual's opinion at operational time $\tau$ is well defined. Introducing the discrete time interval $\Delta\tau$ after $n$ ticks of the individual's clock, results in

$$Q_n = (1 - \lambda_0 \Delta\tau)^n Q_0. \qquad (5.34)$$

The solution to the discrete equation (5.34) is an exponential, in the limit $\lambda_0 \Delta\tau \ll 1$, such that $n \to \infty$ and $\Delta\tau \to 0$, but in such a way that the product $n\Delta\tau$ becomes the continuous operational time $\tau$. However when the individual is part of a dynamic network the exponential no longer describes her behavior. Adopting the subordination interpretation, the discrete index $n$ is an individual's operational time, which is stochastically connected to the chronological time $t$, in which the global behavior of the network is observed. So what is the behavior of the individual in chronological time? The answer to that question depends on the network's dynamics.

The dynamic network we adopt for this discussion is that generated by the decision making model (DMM), which is a member of the Ising universality class. Being a member of the Ising universality class means that the solution to the DMM equations of motion has the same scaling behavior as does the Ising model for phase transitions in non-equilibrium statistical physics [356]. In the DMM an isolated individual randomly switches, back and forth, between two states, according to the discrete operational time. The subordination argument, which is presented below, is applied to the individual to quantify the influence of the network on the individual's dynamics and the discrete index becomes tied to the occurrence of fluctuating events generated by the network's dynamics [359]. The properties of these network fluctuations become strongly amplified as the DMM control parameter approaches its critical value and the network dynamics undergoes a phase transition.

Adopting the subordination interpretation, we define the discrete index $n$ as the operational time that is stochastically connected to the chronological time $t$ in which the global behavior is observed. We assume that the chronological time lies in the interval $(n-1)\Delta\tau \le t \le n\Delta\tau$, and consequently, the equation for the average dynamics of the individual is given by [253]:

$$\langle Q(t) \rangle = \sum_{n=0}^{\infty} \int_0^t G_n(t,t') Q_n dt', \qquad (5.35)$$

where we have the average response of a typical individual to the ensemble of network events.

The physical meaning of equation (5.35) is determined by considering each tick of the internal clock $n$, as measured in experimental time, to be an event induced by the network. Since the observation is made in experimental time, the time intervals between events define a set of independent, identically distributed, (iid) random variables. The integral in equation (5.35) is then built up according to renewal theory [334]. After the $n$-th event, the individual changes from state $Q_{n-1}$ to $Q_n$, where she remains until the next event is generated. The sum over $n$ takes into account the possibility that any number of events could have occurred prior to an observation at experimental time $t$, and $G_n(t,t')dt'$ is the probability the last event occurs in the time interval $(t', t' + dt')$.

### 5.2.2 Network survival

We assume that the waiting times between consecutive events in equation (5.35) are *iid* random variables, so that the kernel is defined

$$G_n(t,t') = \Psi(t-t')\psi_n(t'). \tag{5.36}$$

The probability that no event has occurred in a time $t$, since the last event, is given by the survival probability $\Psi(t)$. This nomenclature comes from the theory of failure, where an event is a failure mode, so no event occurring implies survival. In addition a process being renewal means that events are independent of one another. Individual events occur statistically with a waiting-time PDF $\psi(t)$ and taking advantage of their renewal nature, the waiting-time PDF for the $n$-th event in a sequence is connected to the previous event by a convolution integral

$$\psi_n(t) = \int_0^t dt' \psi_{n-1}(t')\psi(t-t'), \tag{5.37}$$

and $\psi_0(t) = \psi(t)$. The waiting-time PDF is related to the survival probability through:

$$\psi(t) = -\frac{d}{dt}\Psi(t). \tag{5.38}$$

The intermittent statistics of network events were shown by direct numerical integration of the DMM two-state master equation (TSME) [320] to have an IPL waiting-time PDF. Here, we select the probability of no event occurring up to time $t$ to be a generalized hyperbolic PDF:

$$\Psi(t) = \left(\frac{T}{T+t}\right)^{\alpha}, \tag{5.39}$$

which is asymptotically an IPL. Consequently, the waiting-time PDF is renewal, as well as being IPL. As mentioned, the rationale for choosing equation (5.39) for the survival probability is the temporal complexity observed in numerical calculation of the DMM. This survival probability has been numerically determined using the DMM for networks with 2500 individuals, as depicted in Figure 5.5. The elements in the DMM consist of two-state oscillators with the interaction between elements determined by an $N$-dimensional TSME.

To find an analytical expression for the behavior of the individual in experimental time, it is convenient to study the Laplace transform of equation (5.35):

$$\langle \widehat{Q}(u) \rangle = \widehat{\Psi}(u) \sum_{n=0}^{\infty} Q_n [\widehat{\psi}(u)]^n, \tag{5.40}$$

where the kernel was replaced by equation (5.36). The renewal relation for the waiting-time PDF yields

$$\widehat{\psi}_n(u) = [\widehat{\psi}(u)]^n,$$

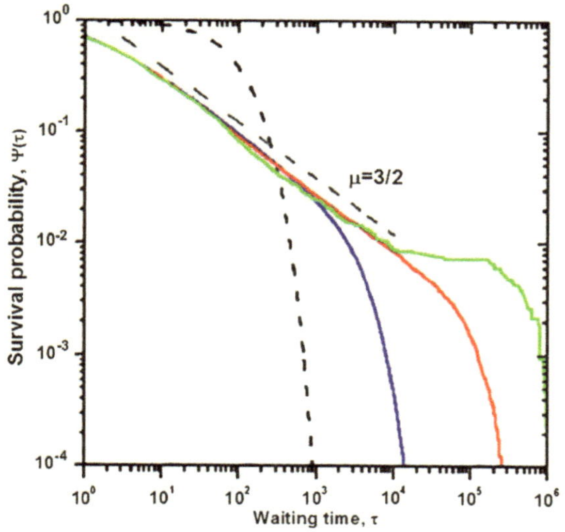


**Figure 5.5:** Simulations of the survival probability for the global variable of a DMM were performed on a lattice of size $N = 50 \times 50$ nodes, with periodic boundary conditions for $g_0 = 0.01$ and increasing values of the control parameter $K$. Blue, red, and green lines correspond to $K = 1.50, 1.70$, and 1.90, respectively. The critical value of the control parameter is $K_c \approx 1.72$. The black dashed line on the plots of $\Psi(\tau)$ denotes an exponential distribution, with the decay rate $g_0$. The gray dashed line denotes an inverse power law function with exponent $\mu - 1$. (From [321] with permission.)

which is used and results from the convolution structure of equation (5.37). With the operational discrete time solution equation (5.34) inserted into equation (5.40) we obtain

$$\langle \widehat{Q}(u) \rangle = \widehat{\Psi}(u) \sum_{n=0}^{\infty} (1 - \lambda_0 \Delta\tau)^n [\widehat{\psi}(u)]^n Q_0. \tag{5.41}$$

Performing the sum and noting the relationship given by the inverse Laplace transform of equation (5.38), results in

$$\langle \widehat{Q}(u) \rangle = \frac{1 - \widehat{\psi}(u)}{u} \frac{1}{1 - (1 - \lambda_0 \Delta\tau)\widehat{\psi}(u)} Q_0. \tag{5.42}$$

This last equation can be expressed in the form:

$$\langle \widehat{Q}(u) \rangle = \frac{1}{u + \lambda_0 \Delta\tau \widehat{\Phi}(u)} Q_0, \tag{5.43}$$

with the newly defined function

$$\widehat{\Phi}(u) = \frac{u\widehat{\psi}(u)}{1 - \widehat{\psi}(u)}, \tag{5.44}$$

which is the Montroll–Weiss memory kernel obtained in the continuous time random walk model (CTRW) [223]. In the asymptotic limit $u \to 0$, if the survival probability is given by equation (5.39) the solution to the FDE in Laplace variables, given by equation (5.43) simplifies to

$$\langle \widehat{Q}(u)\rangle = \frac{1}{u+\lambda u^{1-\alpha}} Q_0 \tag{5.45}$$

with the parameter value:

$$\lambda = \frac{\lambda_0 \Delta\tau}{T\Gamma(2-\alpha)}. \tag{5.46}$$

It is readily shown that

$$\mathcal{LT}^{-1}\left\{\frac{u^{\alpha-1}}{u^{\alpha}+\lambda};t\right\} = \sum_{k=0}^{\infty} \mathcal{LT}^{-1}\left\{\frac{1}{u}\left(\frac{-\lambda}{u^{\alpha}}\right)^{k};t\right\} = \sum_{k=0}^{\infty} \frac{(-\lambda t^{\alpha})^{k}}{\Gamma(k\alpha+1)}$$

so that equation (5.45) has an inverse Laplace transform solution that is a MLF.

Consequently, the subordination process results in the ordinary rate equation, through the inverse Laplace transform of equation (5.45), being replaced with the fractional rate equation [307, 311]:

$$D_t^{\alpha}[\langle Q(t)\rangle] - \frac{t^{-\alpha}}{\Gamma(1-\alpha)} Q_0 = -\lambda\langle Q(t)\rangle, \tag{5.47}$$

where we have introduced the RL fractional operator, defined earlier. Of course, we could just as well have defined the dynamic equation (5.47) in terms of the Caputo or MRL fractional derivatives to obtain

$$\partial_t^{\alpha}[\langle Q(t)\rangle] = -\lambda\langle Q(t)\rangle, \tag{5.48}$$

in which the initial condition has been absorbed into the definition of the fractional derivative. Note that these fractional rate equations reduce to the ordinary rate equation in the limit $\alpha = 1$. Consequently, the solution to equation (5.47) must reduce to the exponential in this limit, as does the solution to equation (5.48).

The DMM network dynamics are known to undergo a phase transition as a control parameter is varied and the results of the calculation in the subcritical, critical, and supercritical regimes of the DMM dynamics, are recorded for a single person. This randomly chosen individual was used in the evaluation of the probability for the single individual interaction with the other members of the network. The calculations were done on an $N = 100 \times 100$ node, two-dimensional lattice, with nearest-neighbor interactions. The time-dependent average solution calculated over an ensemble of randomly chosen individuals is depicted in Figure 5.6, where the average is taken over $10^4$ independent realizations of the dynamics. The analytic solution is obtained from equation (5.48) to be a Mittag-Leffler function.

The rate parameter in the MLF is fit to the numerical calculations of a two dimensional lattice network of 10 000 individuals restricted to nearest neighbor interactions. Note that the influence of the other 9 999 members of the highly nonlinear complex dynamic network on the individual of interest is here predicted by the solution of a linear fractional differential equation without linearizing the dynamics. The response

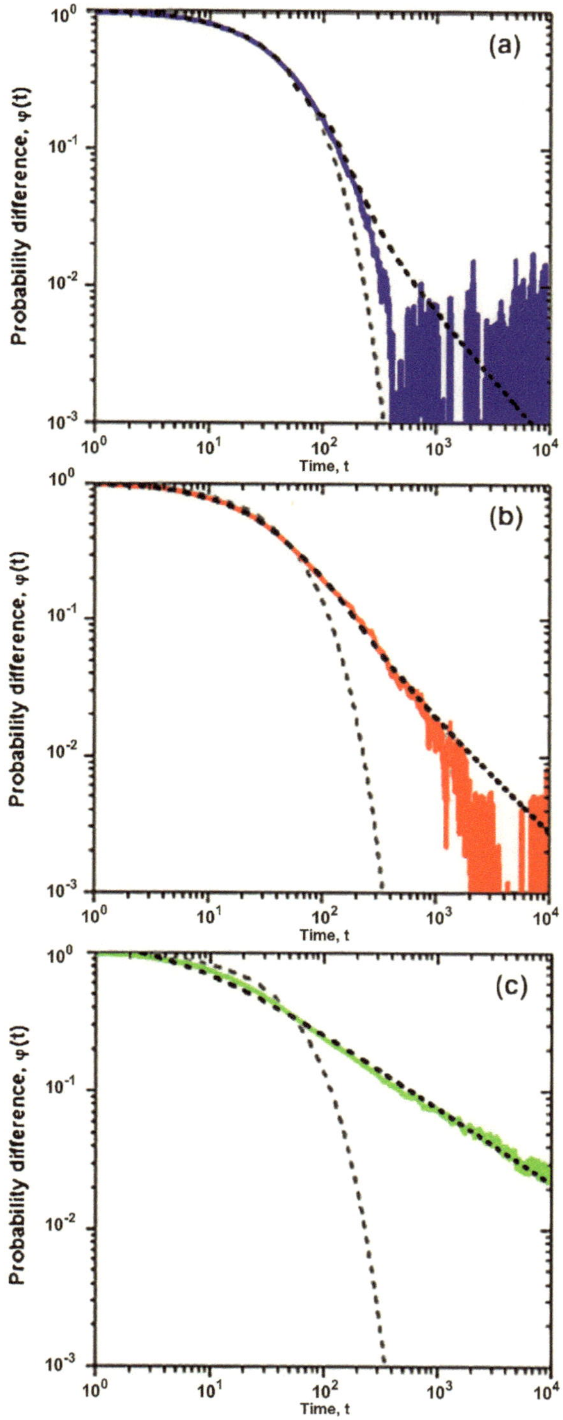


**Figure 5.6:** The dashed curves are the exponentials for the average opinion of an isolated individual. The dotted curves are the fits of the MLF to the erratic network dynamics calculated using the DMM for a network of $10^4$ elements: (a) subcritical; (b) critical; (c) supercritical. (From [358] with permission.)

of the individual to the group, mimics the group's behavior most closely when the control parameter is equal to or greater than the critical value. Note the behavior of the isolated individual depicted by the short-dashed curve in the figure.

In other words, in the DMM, an isolated person with Poisson statistics, becomes an interactive person with Mittag-Leffler statistics. In this way an individual is transformed by those with whom s/he interacts. This transformation of an individual's behavior need not be that of a person, but could be the herding of animals, the flocking of birds, or the swarming of insects. In an allometry context, this change in behavior of the functionality of an individual is a consequence of the complexity of the collective dynamics. Said differently, the average functionality of the individual, here that functionality is related to the probability of the individual switching states, is determined by the average complexity of the group to which that individual belongs. Here that complexity is related to the probability that the entire group switches states at one time. The individual is responding to the dynamic complexity of the group and their individuality is replaced with the more acceptable behavior of the group.

The numerical solution to the nonlinear equations of the DMM given by the millions of individual calculations is shown in Figure 5.6 to be replicated by the analytic solution to the linear fractional rate equation. This suggests that the scaling behavior of the complexity measures manifest in the AR, might be captured by the fractional calculus, as we subsequently show.

The subordination argument more generally yields a fractional Langevin equation (FLE) in chronological time [359] in terms of the RL fractional derivative:

$$D_t^{\alpha}[Q(t)] - \frac{t^{-\alpha}}{\Gamma(1-\alpha)} Q_0 = -\lambda Q(t) + f(t), \tag{5.49}$$

or in terms of the Caputo or MRL fractional derivative

$$\partial_t^{\alpha}[Q(t)] = -\lambda Q(t) + f(t), \tag{5.50}$$

either of which is satisfactory. The fractional index $\alpha$ is determined by the IPL survival probability of the DMM network, which determines how long an observer must wait, between the occurrence of successive changes of global opinion. The zero-centered fluctuations of the random force $f(t)$ are determined by the statistics entailed by the finite size of the complex network. Consequently, the ensemble average of equation (5.49) and (5.50) yields equation (5.47) and (5.48), respectively.

### 5.2.3 Network control

As pointed out by Liu et al. [189], the ultimate understanding of complex networks is reflected in the ability to control them. Recent observations of the interconnectedness of infrastructure networks [52], facilitating the spread of failures [54], or the tight coupling between banking institutions, posing danger to the stability of global financial systems [26], demonstrate the importance of developing a systematic approach to influence and/or control of complex networks. The analysis presented in this chapter offers an alternative to the attempt at addressing this need by trying to impose the conditions of traditional control theory [44] onto the dynamics of complex networks.

To illustrate this point we consider a control that mimics the stochastic properties of the model: the behavior of an individual is virtually unaffected, while the network as a whole tracks the control signal. This is achieved by a slight modification of the transition rates for a small, randomly selected percentage of the individuals. In particular, the transition rate of an affected node is modified by a control signal $\upsilon(t)$ that is a square wave with period $\Omega$ much larger than the characteristic time scale of the single individual. The additional factor can be thought as a fifth node that interacts with an individual person.

Using the *fifth node* control it is not possible to determine numerically whether the network is being controlled solely by observing the dynamics of single individuals. As depicted in Figure 5.7 (c), the survival probability function for a single individual interacting with the original network, compared to one including 3% of the network elements perceiving the control, is indistinguishable. However, the global behavior of the network undergoes pronounced change once the control is applied to even this small percentage of the elements. The temporal variability of the network's average global variable $\xi(K,t)$, characterized by an IPL survival probability $\Psi(\tau)$, prior to applying the control signal, becomes more regular (Figure 5.7 (b)), adopting the control time scale $\Omega$ (Figure 5.7 (d)).

Finally, we use the linear cross-correlation $\chi(K)$ between the control signal $\upsilon(t)$ and the global variable $\xi(K,t)$ to quantify the effect of the control on the network. Its value for increasing percentages of randomly chosen individuals perceiving control is shown in Figure 5.8. The peaking of the cross-correlation at particular value of coupling parameter $K$ indicates that the control variable has maximum influence when the network time scale defined by the survival probability of the global variable is of the same order of magnitude as the period of the control $\Omega$.

## 5.3 Fractional Poisson process

Up to this point we have presented two interpretations of the extension of the dynamics of complex phenomena to fractional differential equations, using either a distribution of rates, or the method of subordination. The dynamics of a Poisson process, since it is also written as the solution to a set of differential equations, can be generalized to the fractional case. This generalization produces a new statistical process, called the fractional Poisson process (FPP). We discuss this new statistical distribution here for three reasons: first, the FPP provides some added experience in solving fractional rate equations; second it has some properties that provide insight into the dynamics

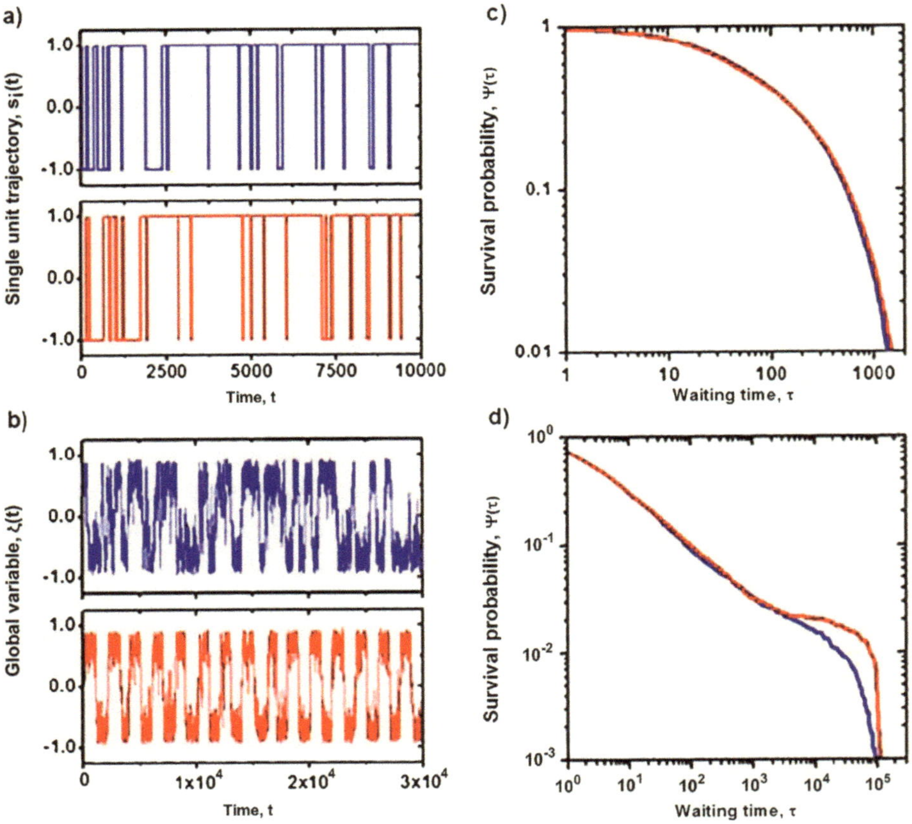


**Figure 5.7:** Comparison of the temporal evolution and corresponding survival probability for the behavior of a single node and the global variable without (blue) and with control (red). Numerical simulations were performed on a lattice of $N = 20 \times 20$ nodes, with $g_0 = 0.01$ and $K = 1.7$. Three percent of randomly selected nodes of the lattice were affected by a square wave with the period $\Omega = 2 \times 10^5$. (From [322] with permission.)

of complex phenomena; finally, our purpose is to determine what complexity entails about the behavior of allometry phenomena.

### 5.3.1 Poisson process dynamics

Following, in part, Laskin's discussion of Poisson processes [179], the properly normalized probability for $n$ events, occurring in a time interval $(0, t)$ is $P(n, t)$, whose discrete dynamics are determined by

$$P(0, t + \Delta t) = P(0, t)(1 - \lambda\Delta t), \tag{5.51}$$

$$P(n, t + \Delta t) = P(n, t)(1 - \lambda\Delta t) + P(n - 1, t)\lambda\Delta t; \quad n \geq 1. \tag{5.52}$$

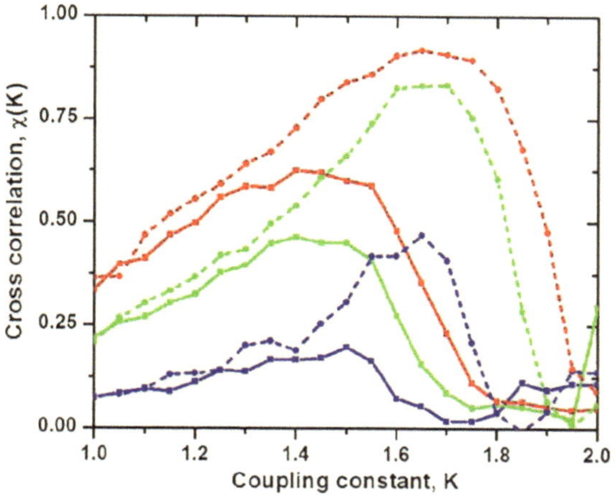


**Figure 5.8:** The cross-correlation of the control signal with the global network dynamics as a function of the coupling parameter K. The control signal is a square wave with period $\Omega = 2 \times 10^4$ and $2 \times 10^5$, denoted by solid and dashed lines, respectively. The percentage of randomly selected individuals being directly coupled to the control increases from 1% (blue), through 3% (green) to 5% (red). The numerical simulations were performed on a lattice of $N = 20 \times 20$ members, with $g_0 = 0.01$. From [322] with permission.

Here the probability of an event occurring in a time interval $\Delta t$ is $\lambda\Delta t$, so that equation (5.51) is the probability that no event occurs up to time $t + \Delta t$, which is the product of the probability that no event occurs up to time $t$ and the probability that no event occurs in the time interval between $t$ and $t + \Delta t$. On the other hand, equation (5.52) denotes the probability of $n$ events occurring in time $t + \Delta t$, to be determined by the sum of the probability that $n$ events occur in the time $t$, but no event occurs in the subsequent interval $\Delta t$, and $n - 1$ events occur in the time $t$, with one event occurring in the subsequent interval $\Delta t$. Consequently, rearranging terms in the latter equation yields

$$\frac{P(n, t + \Delta t) - P(n, t)}{\Delta t} = -\lambda[P(n, t) - P(n-1, t)]; \quad n \geq 1, \tag{5.53}$$

which allows us to construct the first-order rate equations in the limit $\Delta t \to 0$:

$$\frac{\partial P(0, t)}{\partial t} = -\lambda P(0, t), \tag{5.54}$$

$$\frac{\partial P(n, t)}{\partial t} = -\lambda[P(n, t) - P(n-1, t)]; \quad n \geq 1, \tag{5.55}$$

with the initial condition given by $P(n, t = 0) = P_0(n)$. Equations (5.54) and (5.55), taken together, constitute the dynamics of a traditional Poisson process.

The solution to the system of equations describing the dynamics of a Poisson process can be obtained using the probability generating function [179]

$$G(s,t) = \sum_{n=0}^{\infty} s^n P(n,t), \tag{5.56}$$

whose dynamics are obtained by multiplying equation (5.55) by $s^n$, summing over the number of events $n$, and relabeling terms in the series to obtain

$$\frac{\partial G(s,t)}{\partial t} = -\lambda(1-s)G(s,t). \tag{5.57}$$

The resulting initial value problem defined by equation (5.57), with $G(s,0) = 1$, has the solution for the probability generating function

$$G(s,t) = \exp[-\lambda(1-s)t], \tag{5.58}$$

which can be used to determine the probability:

$$P(n,t) = \frac{1}{n!} \frac{\partial^n G(s,t)}{\partial s^n}\bigg|_{s=0}. \tag{5.59}$$

Inserting equation (5.58) into equation (5.59) and preforming the indicated operations yields

$$P(n,t) = \frac{(\lambda t)^n}{n!} e^{-\lambda t}, \tag{5.60}$$

so that the number of events occurring in the time $t$ has a Poisson distribution, with the mean number of events given by $\lambda t$.

To generalize this procedure for determining the probability to the fractional case, we use the notion of subordination, which again implies the existence of two distinct times [311]. First, is the operational time $\tau$, which is the internal time, determining the generation of a Poisson process by means of the ordinary differential equation (ODE) of a non-fractional system, such as given by equation (5.54). Second, is chronological time $t$, which is measured by the clock of an external observer. The subordination procedure transforms the ODE in operational time into a FDE in chronological time, as we demonstrate for the FPP.

### 5.3.2 Fractional Poisson dynamics

The behavior of an isolated individual, as we found using the DMM, does not change dramatically as she begins to interact with a group, but becomes less individualistic and more herd-like as the strength of the interaction with the other members of the network increases. The form of behavior modification is strongly dependent on the critical value of the control parameter, which determines the transition to collective behavior, but is relatively insensitive to the size of the group. The individual's Poisson nature is not lost quickly, but the change reveals an adaptation on the part of the individual to the collective influence of the others within the group. This behavior was previously investigated by [361], who examined only the change in the survival probability, that being the $n = 0$ term of the Poisson distribution, for subcritical, critical, and supercritical dynamics. This result was discussed in the previous section.

We here adapt the subordination method, previously applied to the survival probability, to the dynamics of the full Poisson probability and generalize equations (5.54) and (5.55) to fractional dynamics, using subordination reasoning [305, 311]. Typically, in the operational time frame, the temporal behavior of an individual is regular and evolves exactly according to the ticks of a clock leading to the Poisson distribution. Therefore it is assumed that the trajectory of a network's individual in operational time is well defined and given by $P(n,\tau)$. In operational time an individual's behavior appears regular, but to an experimenter, observing the elements from outside the network, we assume that their temporal behavior appears erratic, as discussed in Section 5.2. Because of the random nature of the chronological time evolution of the individual, the subordination process involves an ensemble average over many realizations of the individual's dynamics, each person evolving according to their own internal clock.

The details of subordination in this case are relegated to the Appendix in order not to obscure the argument with too many details, it is a modest extension of the arguments presented in Section 5.2. The probability is shown to be given by

$$\frac{\partial \mathcal{P}(n,t)}{\partial t} = -\lambda \int_0^t \Phi(t-t')\{\mathcal{P}(n,t') - \mathcal{P}(n-1,t')\}dt', \tag{5.61}$$

a generalized master equation (GME) for the probability of the occurrence of $n \geq 1$ events in the time $t$, with the probability being given by

$$\mathcal{P}(n,t) \equiv \langle P(n,t) \rangle. \tag{5.62}$$

The brackets denote an average over the statistical distribution of subordinated ticks of the operational clock. When the memory kernel is a delta function in time

$$\Phi(t) = \delta(t), \tag{5.63}$$

the GME reduces to a differential equation for the Poisson process given by equation (5.55). Consequently, the subordination process induces a memory into the dynamics, which is determined by the memory kernel.

The dynamics of the probability using the empirically determined survival probability from the DMM numerical calculation [361] is presented in the Appendix. The fractional differential equation for the probability generating function obtained in this way is

$$\partial_t^\alpha [\mathcal{G}_\alpha(s,t)] = -(1-s)\gamma^\alpha \mathcal{G}_\alpha(s,t), \tag{5.64}$$

in terms of the Caputo fractional derivative. The initial value solution to this fractional rate equation for the probability generating function equation (5.64) is

$$\mathcal{G}_\alpha(s,t) = E_\alpha(-(1-s)\gamma^\alpha t^\alpha), \tag{5.65}$$

where the MLF is defined by equation (5.31).

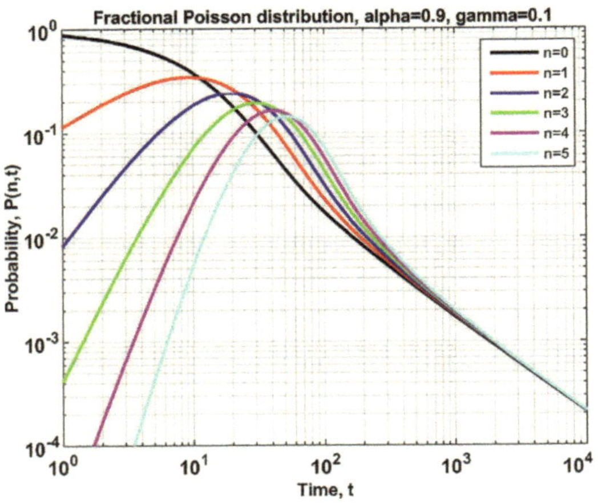


**Figure 5.9:** The fractional Poisson process is graphed as a function of time for six values of the number of events and $\alpha = 0.90, \gamma = 0.1$. The $n = 0$ curve (black) is the survival probability for the FPP. Thanks to M. Turalska for carrying out these numerical calculations using equation (5.66).

The probability is, using equation (5.59), determined from the derivatives to the probability generating function to be [179]

$$\mathcal{P}_\alpha(n,t) = \frac{(-z)^n}{n!} E_\alpha^{(n)}(z)\bigg|_{z=-\gamma^\alpha t^\alpha}, \tag{5.66}$$

where the superscript on the MLF denotes the $n$-th derivative with respect to its argument, and then inserting the indicated value of the independent variable. Equation (5.66) is the solution to the set of FPP rate equations

$$\partial_t^\alpha[\mathcal{P}_\alpha(0,t)] = -\gamma^\alpha \mathcal{P}_\alpha(0,t) \tag{5.67}$$

$$\partial_t^\alpha[\mathcal{P}_\alpha(n,t)] = -\gamma^\alpha[\mathcal{P}_\alpha(n,t) - \mathcal{P}_\alpha(n-1,t)]; \quad n \geq 1, \tag{5.68}$$

which are direct generalizations of equations (5.54) and (5.55), respectively, with $\lambda$ replaced with $\gamma^\alpha$. Note that equation (5.67) had been previously derived from the DMM calculation to denote the fractional dynamics of the network survival probability [359, 361].

The FPPs for six values of the number of events are graphed in Figure 5.9 for the fractional derivative of order $\alpha = 0.90$. Note that the FPP has an asymptotic IPL in time rather than the characteristic exponential of the traditional Poisson process. The IPL dominates the FPP at times greater than $1/\gamma$.

The explicit series form for the probability for the occurrence of $n$ events in the time $t$ can be constructed by inserting the series expansion for the MLF into equation (5.66) and evaluating the derivatives to obtain

$$\mathcal{P}_\alpha(n,t) = \sum_{k=n}^{\infty} \binom{k}{n} (-1)^{n-k} \frac{(\gamma t)^{k\alpha}}{\Gamma(k\alpha+1)}. \tag{5.69}$$

Laskin [179], derived this expression using equation (5.66). The same series was obtained using other techniques by Meerschaert et al. [215], as well as, Mainardi et al. [197]. The latter were the first to label this process FPP.

Recall the influence of a large network on the behavior of an individual depicted in Figure 5.6. In that figure the lowest-order term in the FPP, that being the probability of no event occurring in the time interval $(0, t)$, compared favorably with the numerical calculations of the DMM. The excellent fit was made for the subcritical, critical and supercritical values of the control parameter. It was therefore interesting to determine that the higher-oder terms in the FPP do not do as well in capturing the statistics of the DMM complex network in the tails of the distribution. Consequently, the response of the individual to the behavior of the group may not be as simple as found using DMM.

## 5.4 A closer look at complexity

The subordination argument has provided some insight into one source of fractional operators and suggests at least one way to quantify the complexity of a phenomenon using the fractional calculus. We have seen that the phase transition, or the onset of collective behavior in group decision making, is associated with the complexity generated by nonlinear dynamics, but it is probably time to take a closer look as to what goes into the making of complexity.

Complexity as a science is transdisciplinary, so it might be argued that the schema constructed for its understanding, based on the physics paradigm, are incomplete. On the other hand, many believe that the principle of objectiveness, namely the objective existence of mechanism-based laws, gives physics an important advantage in addressing the difficult task of understanding complexity. Here we notice the hierarchy of science from the "hard" to the "soft", emerges from adopting a materialistic-objective point of view, which is closely related to reductionism [67]. The physics-based approach is not in conflict with the "holistic" theories of complexity, but the former approach is less flexible than the latter.

An experiment often consists of studying the response of a complex network to controlled external perturbations. This set of responses constitute the information the observer extracts from the network and constitutes what the experimenter can know. The attendant difficulty encountered in understanding, controlling, or predicting these responses, is intuitively interpreted as complexity. The greater the difficulty, the greater the level of complexity.

It is useful to list the properties associated with the complexity of a system, because we are seeking quantitative measures that include an ordinal relation for complexity, that is, one system can be twice as complex as another system. We note, however, that in everyday usage, phenomena with complicated and intricate features, having both the characteristics of randomness and order are called complex. We adopted this working definition of complexity, because it served our purpose and there is no consensus, among scientists, poets or philosophers, on what constitutes a good quantitative measure of complexity. Therefore, any list of traits of complexity is arbitrary and idiosyncratic, but the following list contains traits that are part of its complete characterization [101, 329]:

- A complex system typically contains many elements. As the number of elements increases so too does the system's complexity up to a certain point.
- A complex system typically contains a large number of relations among its elements. These relations usually constitute the number of independent dynamical equations.
- The relations among the elements are generally nonlinear in nature, often being of a threshold, or saturation character, or more simply of a coupled, deterministic, nonlinear dynamical form. The system often uses these relations to evolve in a self-regulating way.
- The relations among the elements of the system are constrained by the environment and often take the form of being externally driven or having a time-dependent coupling. This coupling is a way for the system to probe the environment and adapt its responses for maximal likelihood of survival.
- A complex system often remembers its evolution for a long time and is therefore able to adapt its behavior to changes in internal and external conditions.
- A complex system is typically a composite of order and randomness, but with neither being dominant.
- Complex systems often exhibit scaling behavior, over a wide range of time and/or length scales, indicating that no one scale is able to characterize the evolution of the system.

These are among the most common properties identified to characterize complex systems, see, for example, [75], and in a set of dynamical equations, these properties can often be kept under control by one or more parameters. The values of these parameters can sometimes even be taken as measures for the complexity of the system. Note the use of qualifiers, such as often, typically, and usually in the list. Not all the properties listed are present all the time, and the qualifiers enable us to extend the list of properties, without fixing the definition of complexity so tightly that it chokes off applications. This way of proceeding is, however, model-dependent and does not always allow for comparison, between the complexities of distinctly different phenomena, or more precisely between distinctly different models of phenomena.

In the above list we included one of the most subtle concepts in our discussion of complexity; the existence and role of randomness [12, 246]. Randomness is associated with our inability to predict the outcome of a process, such as flipping a coin, rolling dice, or drawing a card from a deck of cards. It also applies to more complicated phenomena, for example, we cannot know, with certainty, the outcome of an athletic contest, such as a basketball or football game. More profoundly, we cannot say, with certainty, what will be the outcome of a medical operation, such as the removal of a cancerous tumor. From one perspective, the unknowability of such events has to do with the large number of elements in the system, so many in fact, that the behavior of the system ceases to be predictable [194]. On the other hand, we now know that having only a few dynamical elements in the system does not insure predictability or knowability. It has been demonstrated that the irregular time series observed in such disciplines as economics, chemical kinetics, physics, logic, physiology, biology and on and on, are at least, in part, due to chaos [180].

Technically, chaos is a sensitive dependence on initial conditions of the solution to a set of nonlinear, deterministic, dynamical equations. Practically, chaos means that the solutions to such equations look erratic and may pass all the traditional tests for randomness, even though they are deterministic. Therefore, if we think of random time series as complex, then the output of a chaotic generator is complex. However, we know that something as simple as a one-dimensional, quadratic map can generate a chaotic sequence. Thus, using the common notion of complexity, it would appear that chaos implies the generation of complexity from simplicity. This is part of Poincaré's legacy of paradox. Another part of that legacy is the fact that chaos is a generic property of nonlinear dynamical systems, which is to say, chaos is ubiquitous; all systems change over time, and because they are nonlinear, they manifest chaotic behavior, over varying scales of time; from the microscopic to the cosmological.

A nonlinear system, with only a few dynamical variables, has chaotic solutions and therefore can generate random patterns. So we encounter the same restrictions on our ability to know and understand a system, when there are only a few dynamical elements, as when there are a great many dynamical elements, but for very different reasons. Let us refer to the former random process as *noise*, the unpredictable influence of the environment on the system of interest. Here the environment is assumed to have an infinite number of elements, all of which we do not know, but they are coupled to the system of interest and perturb it in a random, that is, unpredictable way [184, 389]. By way of contrast, chaos is an implicit property of a complex system, whereas noise is an explicit property of the environment in contact with the system of interest. Chaos can therefore be controlled and predicted over short time intervals, whereas noise can, neither be predicted, nor controlled, except perhaps through the way it is allowed to interact with the system.

The distinction between chaos and noise, made here, highlights one of the difficulties of formulating an unambiguous measure of complexity. Since noise cannot be predicted, or controlled, it might be viewed as being simple. Thus, systems with many degrees of freedom, which manifest randomness are considered simple. On the other hand, a nonlinear dynamic system with only a few dynamical elements, when it is chaotic, might also be considered simple. In this way the idea of complexity is ill-posed and a new approach to its definition is required.

In early papers on systems theory it was argued that the increasing complexity of an evolving system can reach a threshold, where the system is so complicated that it is impossible to follow the dynamics of the individual elements [329]. At this point new properties often emerge and the new organization undergoes a completely different type of dynamics; for example, the transition of fluids from laminar to turbulent flow, or the phase transition of gas to liquid. The details of the interactions among the individual elements are substantially less important than is the "structure", the geometrical pattern, of the new aggregate. Increasing the number of elements further, or alternatively increasing the number of relations, often leads to a complete "disorganization" and the stochastic approach becomes a better description of the system's behavior. If randomness (noise) is to be considered as something simple, as it is intuitively, one has to seek a measure of complexity that decreases in magnitude in the limit of the system having an infinite number of elements. This reasoning leads to the inverted parabola-like measure of complexity sketched in Figure 1.1.

### 5.4.1 Entropies

Historically thermodynamics was the first discipline in physics to systematically investigate the order and randomness of complex systems, since it was here that the natural tendency of things to become disordered was first quantified. As remarked by Schrödinger in his groundbreaking work, *What is Life*? [290]:

> The non-physicist finds it hard to believe that really the ordinary laws of physics, which he regards as prototype of inviolable precision, should be based on the statistical tendency of matter to go over into disorder.

In this context the quantitative measure of "disorder", which has proven to be very valuable, is entropy and thermodynamic equilibrium is the state of maximum entropy. Of course, since entropy has been used as a measure of disorder, it seems that it should also be useful as a measure of complexity. If living matter is considered to be among the most complex of systems, for example the human brain, then it is useful to understand how the enigmatic state of being alive is related to entropy. Schrödinger maintained that a living organism can only hold off the state of maximum entropy, that being death, by absorbing negative entropy, or *negentropy*, from the environment. He points out that the essential thing in metabolism is organisms must free themselves from the entropy they produce in the process of living.

We associate complexity with disorder, which is to say with limited knowability, and order with simplicity, or absolute knowability. This rather comfortable separation into the complex and the simple, or the knowable and the unknowable, in the physical sciences breaks down, once a rigorous definition of entropy is adopted and applied outside the restricted domain of thermodynamics. It becomes apparent, because of the fundamental ambiguity in the definition of complexity, that even adopting the single concept of entropy as the measure of complexity, leads to multiple definitions of entropy that are not always consistent with one another.

Following Cambel [59] we can divide entropies, which have been discussed historically, into three rough-hewn categories: the macroscopic, the statistical and the dynamical. In the first group we find the entropy stemming from thermodynamics, for example, the original macroscopic S-function entropy of Clausius, as used by Boltzmann [40] and subsequently by Prigogine [256]. The second category contains the entropy resulting from the assumption that there exists a PDF to characterize an ensemble of realizations of the system, such as was first introduced by Gibbs [113]. Here the activity on the microscale, or the dynamics of the individual elements in phase space, are related to what occurs macroscopically, or at the system level, through the average PDF. A special role is enjoyed by the statistical entropy, which was chosen to quantify information by Wiener [377], as well as Shannon [294], which relates the statistical behavior of a system to the concept of information. Finally, the dynamical entropy, such as the one introduced by Kolmogorov [171], are derived from the geometry of the system's dynamics in phase space. Other possible choices for categories would have to demonstrate a distinct advantage over those listed above and would overlap with the chosen categories in various ways.

### 5.4.2 Information, incompleteness and uncomputability

In addition to the arguments given above, there might also exist other reasons why, given our present state of knowledge, physical theories do not provide a satisfactory description of reality. It is not sufficient for physics to describe the world within the laboratory, it must also faithfully describe the world in which we live. It seems clear that reductionism is not sufficient to describe systems, where the patterns of information flow often play roles that are more important than those of micro-dynamics, for example, the phase transitions to consensus in social groups. However, the macrodynamic rules still need to be consistent with the micro-dynamic rules. If new properties emerge, even if it is impossible in practice to predict them from micro-dynamics, they must be implicit in the micro-variable equations. This weak reductionistic assumption is part of the objectivity principle.

Our understanding of global behavior is achieved from a holistic point of view, but the emerging behaviors need to be compatible with a weakly reductionistic perspective. Otherwise we would be tempted to imagine different laws for different space and/or time scales, or different levels of complexity. This, in turn, inhibits any possible mechanistic (objective) view of reality. We stress that in our perspective the principle of objectivity, does not necessarily mean that the laws are deterministic, but a seed of randomness may be involved. Actually we believe that a seed of randomness must be involved in any fundamental description of reality.

The English physicist Penrose [244], stressed another way in which standard physical theories fail to describe reality. He developed an extended argument devoted to showing the impossibility of creating an artificial intelligence, using standard computers. In his discussion, he explains how physics is basically "computable", which is to say that the laws of physics can be faithfully implemented using computer programs, and cannot therefore explain cognitive activity. Many scientists take the position that awareness and consciousness require properties that computers lack, see, for example, [192].

Penrose establishes that mathematical reasoning is non-computable, since it is impossible for any computer to have particular mathematical knowledge available to the human brain. The proof of this assertion requires one to define what is meant by a computer, or what is called a "universal Turing machine", as well as, what it can or cannot do, even with unlimited time and memory. Given these constraints, it is possible to use a version of the famous incompleteness theorem of Gödel [124] to prove the assertion, namely that every set of formal mathematical rules is always incomplete. In particular, the knowledge itself of this incompleteness is unavailable to formal theories, but is available to we human beings, and that is because we are able to understand the nature of "paradox". It has been proven that the formal theories can be expressed in terms of computation and vice-versa, so that our capabilities for going beyond what is prescribed for formal theories by Gödel's theorem is a conceptual proof of the existence of non-computable phenomena in the world.

A natural application of computation theory has been to the development of a measure of complexity. This measure can be viewed as a generalization of Shannon's information entropy. It is called *computational complexity* or Kolmogorov–Chaitin complexity [65, 171], after the names of the two mathematicians that independently defined it. This measure applies to binary strings and is defined as the length of the string in bits for the shortest program, which is able to compute the string. Just like entropy, this function reaches a maximum if complete randomness occurs, since genuine randomness is non-computable, one has to specify the entire sequence in the program. The Kolmogorov–Chaitin entropy, like the informational entropy of Shannon, enables one to define conditional, or mutual properties, to establish subadditive properties, that are the common features of complex phenomena. This measure is very useful from a conceptual point of view, but it does not have a practical use, since theorems indicate that it cannot be computed. This particular definition of entropy has been used as a measure of complexity in a number of different fields, including program optimization as well as image and information compression, but it is not useful for us here.

We have argued that the science of complexity is a transdisciplinary approach to the study of reality, not confined to physics, but ranging from biology to economics, and from there to psychology, neurophysiology and the study of brain function. Along that line, Schweber [291] pointed out a crisis generated in physics by the success of RGT:

> The ideas of symmetry breaking, the renormalization group and decoupling suggest a picture of the physical world that is hierarchically layered into quasiautonomous domains, with the ontol-

ogy and dynamics of each layer essentially quasistable and virtually immune to whatever happens in other layers. At the height of its success, the privileged standing of high-energy physics and the reductionism that permeated the field were attacked.

Reductionism was vigorously attacked early on by Anderson [8]. The renormalization group specifies a set of rules for establishing the critical coefficients of phase transition phenomena. Wilson and Kogut [382] prove that the value of these coefficients can be assessed with theoretical models in a way that is totally independent of the detailed nature of elementary interactions. In other words, the RG approach establishes the independence of the different levels of reality, and, even if in principle a person is nothing more than a collection of atoms, individual behavior has to be studied, if ever possible, with scientific paradigms, which do not have anything to do with microscopic dynamics. The leading role of high energy physics in science was based on the implicit assumption that once the fundamental laws of physics are established, all phenomena, at least in principle, can be explained in terms of these laws. The advent of RGT implied that even if a final theory is possible, such as envisioned by Weinberg [333], it cannot be used to address the problems associated with quantifying complexity. On the other hand, this dream of a final theory might also be perceived as a nightmare by people, who like the present author, hope and believe that reality is an inexhaustible source of wonders. We share, on this issue, the same view as Leggett [182]. We believe that the notion of strict determinism must be abandoned and that the settlement of the problem of the great unification in physics, even if it occurs, does not represent the end of physics. Leggett [182] concludes his book with:

> If even a small part of the above speculation is right, then, far from the end of the road being in sight, we are still, after three hundred years, only at the beginning of a long journey along a path whose twists and turns promise to reveal vistas which at the present are beyond our wildest imagination. Personally, I see this as not pessimistic, but a highly optimistic, conclusion. In intellectual endeavor, if nowhere else, it is surely better to travel hopefully than to arrive, and I would like to think that the generation of students now embarking on a career in physics, and their children and their children's children, will grapple with questions at least as intriguing and fundamental as those which fascinate us today-questions which, in all probability, their twentieth-century predecessors did not even have the language to pose.

## 5.5 Recapitulation

In this chapter we took seriously the famous comment made by Einstein that:

> We cannot solve problems by using the same kind of thinking we used when we created them.

We interpret this quote to mean that complex problems require a new way of thinking. A way of thinking quantitatively that does not entail the arcane assumptions of linearity and LFE, but addresses directly the scientific barriers imposed by complexity. This is what we propose to accomplish with the introduction of fractional calculus.

The first barrier addressed, using the fractional calculus approach, was that for the non-differentiability of fractal functions; those used to model phenomena having no characteristic scale. Such non-differentiable functions were shown to have finite fractional derivatives and therefore their dynamics can be described by fractional differential equations. Fractional rate equations were used to describe the relaxation dynamics of viscoelastic materials such as rubber, tar and polymers. The solution to such equations were expressed in terms of a MLF, which is stretched exponential at early times and an IPL at late times. The overlap of the MLF with experimental data is surprisingly good.

The second barrier overcome was establishing that the numerical solution of a large number of nonlinear coupled rate equations in the DMM, could be successfully replaced by the MLF solution to a linear fractional rate equation, with appropriately fit parameters. This replacement demonstrates that the fractional rate equation can faithfully model the lowest-order influence of the complex dynamics of the group on the behavior of the individual as a fractional Langevin equation (FLE). The size of the group determines the strength of the random fluctuations in the FLE and the average interaction with the group is determined by the deterministic part of the FLE. This opens a new trail for research into the origin of ARs, which we follow in the next chapter.

Fractional operators were shown to be a consequence of scaling; whether the scaling was entailed by a distribution of relaxation rates in the underlying process, or the subordination of an individual's operational time to the chronological time of the group. The latter argument was used to demonstrate the change in an individual's behavior from when she is alone to when she interacts with the group. This change in time suggested generalizing the argument to obtain the probability for a FPP, in which the early time probability is that of Poisson, but the late time probability is IPL. The survival probability of the individual, that is the $n = 0$ term, coincides with that of the DMM. However, the higher-order terms do not coincide with the multiple event probabilities, which was not a surprise, because the DMM has a specific kind of interaction that restricts the number of events (changes of state) that can occur within a given time interval. A restriction not imposed by the subordination argument.

The subordination procedure provides an equivalent description of the average dynamics of the single person in terms of a fractional differential equation. The exact solution to this equation determines that the Poisson statistics of the isolated individual becomes Mittag-Leffler statistics due to the interaction of that individual with the complex dynamic network. Consequently the individual's simple random behavior is replaced with one that might serve a more adaptive role in social networks. This adaptation is revealed when control is introduced through a modification of the transition rates and used to directly influence a relatively small number of randomly chosen people. The statistics of the driven individuals do not change perceptively in this latter case, but the DMM network dynamics essentially tracks the control signal for as few as 5% or 20 of the 400 individuals constituting the network being affected.

All this prompted a closer look at complexity and what its potential measures might be like. The major candidate for a complexity measure is entropy; necessitating a brief examination of the macroscopic, statistical and dynamical definitions of entropy. In Chapter 6 we adapt the fractional probability calculus to the derivation of the allometry equations.

## 5.6 Appendix

To determine the average behavior in chronological time, we transform the operational time solution to the chronological time solution adopting the subordination interpretation. We assume that the chronological time lies in the discrete time interval $(n-1)\Delta\tau \leq t \leq n\Delta\tau$ and using a discrete form of equation (5.58), the generating function, obtain

$$\langle G(s,t)\rangle = \sum_{k=0}^{\infty} \int_0^t \Psi(t-t')\psi_k(t') \exp[-\lambda(1-s)k\Delta\tau]dt'. \tag{5.70}$$

It is evident that the generating function resulting from the subordination process inherently involves an ensemble average. Here we see that equation (5.70) replaces the solution to the single person TSME of the DMM, with the transition rates determined by the states dynamically occupied by the other members of the network.

Following the subordination argument in Section 5.2, an analytic expression for the behavior in chronological time can be obtained from the Laplace transform to the generating function

$$\begin{aligned} \widehat{\mathcal{G}}(s,u) \equiv \langle \widehat{G}(s,u)\rangle &= \widehat{\Psi}(u) \sum_{k=0}^{\infty} [\widehat{\psi}(u)]^k [1-\lambda(1-s)\Delta\tau]^k \\ &= \frac{\widehat{\Psi}(u)}{1-[1-\lambda(1-s)\Delta\tau]\widehat{\psi}(u)}. \end{aligned} \tag{5.71}$$

Inserting equation (5.44) into the latter equation yields

$$\widehat{\mathcal{G}}(s,u) = \frac{1}{u+(1-s)\lambda\Delta\tau\widehat{\Phi}(u)}, \tag{5.72}$$

whose inverse Laplace transform yields

$$\frac{\partial \mathcal{G}(s,t)}{\partial t} = -(1-s)\lambda \int_0^t \Phi(t-t')\mathcal{G}(s,t')dt'. \tag{5.73}$$

The derivatives with respect to $s$ commutes with the time integral so that using equation (5.59) in equation (5.73) enables us to write

$$\frac{\partial \mathcal{P}(n,t)}{\partial t} = -\lambda \int_0^t \Phi(t-t')\{\mathcal{P}(n,t')-\mathcal{P}(n-1,t')\}dt', \tag{5.74}$$

a generalized master equation (GME) for the probability of the occurrence of $n$ events in the time $t$, with

$$\mathcal{P}(n,t) \equiv \langle P(n,t) \rangle. \tag{5.75}$$

When the memory kernel is a delta function in time $\Phi(t) = \delta(t)$ the GME reduces to differential equation for the Poisson process given by equation (5.55).

Previous analyses have shown that the global waiting-time PDF is an IPL, so that the asymptotic behavior of an individual in time is determined by

$$\lim_{u\to 0} \widehat{\psi}(u) = 1 - \Gamma(1-\alpha)(uT)^{\alpha}; \quad 0 < \alpha = \mu - 1 < 1 \tag{5.76}$$

so that equation (5.72) reduces to

$$\begin{aligned} \widehat{\mathcal{G}}(s,u) &= \frac{1}{u + (1-s)\gamma^{\alpha} u^{1-\alpha}} \\ &= \frac{u^{\alpha-1}}{u^{\alpha} + (1-s)\gamma^{\alpha}} \end{aligned} \tag{5.77}$$

and the rate parameter has the value

$$\gamma^{\alpha} = \frac{\lambda \Delta\tau}{\Gamma(1-\alpha)T^{\alpha}}. \tag{5.78}$$

The inverse Laplace transform of a reordered form of equation (5.77) yields the fractional rate equation

$$\partial_t^{\alpha}[\mathcal{G}(s,t)] = -(1-s)\gamma^{\alpha}\mathcal{G}(s,t) \tag{5.79}$$

in terms of the Caputo fractional derivatives. The solution to this fractional rate equation is, of course, a MLF.

# 6 Fractional probability calculus

The importance of randomness in characterizing the dynamics of complex systems was discussed in Chapter 4, through the introduction of stochastic forces and random fluctuating parameters. The ideas developed to understand the resulting stochastic differential equations (SDEs), called Langevin equations in the physics literature, were used to generalize the dynamic ontogenetic growth model (OGM). Modeling the uncertainty in the dynamics was accomplished by introducing a stochastic parameter, to obtain a multiplicative noise formulation of the SOGM. The corresponding phase space equation for the associated PDF, was shown to be the Fokker–Planck equation (FPE). The stochastic dynamics of the SOGM was replaced by the FPE, whose asymptotic solution provided a steady-state PDF, with which to characterize the empirical total body mass (TBM) data. This steady-state solution to the FPE gave an excellent fit to data, yielding an asymptotic inverse power-law (IPL) PDF for the TBM. The first moment of the steady state PDF for the interspecies TBM was then shown to yield the interspecies AR. This approach was the first derivation of the empirical AR [355], that is, the AR in terms of average values.

We continue to develop the probability calculus perspective in the present chapter. We do this by inserting memory effects into the kinetics using a fractional generalization of the phase space equations of motion for the PDF. The dramatic complexity of the various allometry systems of interest, both inside and outside the Natural Sciences, are taken into account by extending the application of the fractional calculus to multi-variable fractional equations. The arguments are not intended to be mathematically rigorous, but rely heavily on scientific intuition. However, the demonstrations have enough detail that the interested reader can fill in the gaps with a little effort, or by consulting the cited literature. We begin by demonstrating that the relation between a fractional stochastic differential equation (FSDE) and a fractional Fokker–Planck equation (FFPE), can be obtained using an extension of the arguments developed in Chapter 4 to derive the FPE in fractional form.

However, the ARs are in terms of two system variables, the size and functionality of interest and therefore we develop the fractional phase space equations (FPSEs) for two variables. We sketch one way to extend Zaslavsky's argument for constructing the FPSEs for the ensemble distribution of chaotic trajectories of a single system variable [388] to two variables. The general solution to the potential-free FPSE is shown to have the general form introduced earlier without proof and shown to give rise to the empirical AR.

We find it instructive to introduce Shannon/Wiener information to analyze systems described by the scaling PDF, that is, by the steady-state solution to the FFPE. It is shown that the change in the average value of a system variable is driven by the change in entropy associated with that variable, whether that variable is a measure of the system's functionality or size. An entropy balance requirement entails the em-

DOI 10.1515/9783110535136-006

pirical AR, thereby providing a second perspective from which the empirical AR is derived. We also show that Huxley's original argument 'deriving' the intraspecies AR was not as far from the mark as most investigators have come to believe.

## 6.1 Fractional Fokker–Planck equation

Section 4.4 contains an outline of how to shift perspectives, in terms of formal representations, from the multiplicative SDE for the dynamic variable to the FPE for the PDF. This approach followed the construction given in Lindenberg and West [184]. The procedure was generalized in Chapter 5 to an examination of how to expand the view to include the FSDE for the dynamic variable and from there to adopt the FFPE for the PDF; selected portions of the detailed argument [347] are present here. Consider the fractional generalization of the SDE given in the discussion of the SOGM:

$$\partial_t^{\alpha}[Z_{\xi}(t)] = G(Z_{\xi}) + g(Z_{\xi})\xi(t). \tag{6.1}$$

Note that the formal expression for the phase space distribution is still given by equation (4.73), since the solution to the initial value problem, defined by equation (6.1), is given by $Z_{\xi}(t)$.

The phase space equation for the PDF is constructed by again introducing an arbitrary analytic function $F(z)$, such that its dependence on the stochastic variable of interest is given by the phase space integral

$$F(Z_{\xi}) = \int dz \rho_{\xi}(z,t \mid z_0)F(z). \tag{6.2}$$

The difference in the function between two nearby points in time, is

$$F(Z_{\xi}(t+\Delta t)) - F(Z_{\xi}(t)) = \int dz F(z)[\rho_{\xi}(z,t+\Delta t \mid z_0) - \rho_{\xi}(z,t \mid z_0)],$$

from which we can independently construct the differential limits of both sides of the equation. The first form of the LHS limit is that of the fractional derivative of the function

$$\partial_t^{\alpha}[F] \equiv \lim_{\Delta t \to 0} \frac{F(Z_{\xi}(t+\Delta t)) - F(Z_{\xi}(t))}{\Delta t^{\alpha}}. \tag{6.3}$$

Now, expanding the function in a Taylor series $F(Z_{\xi}(t+\Delta t)) = F(Z_{\xi} + \Delta Z_{\xi})$ about $Z_{\xi}$, since it is analytic in its argument, and retaining the lowest-order term in $\Delta Z_{\xi}$ yields

$$\partial_t^{\alpha}[F] = \lim_{\Delta t \to 0} \frac{dF}{dZ_{\xi}} \frac{\Delta Z_{\xi}}{\Delta t^{\alpha}} = F'(Z_{\xi})\partial_t^{\alpha}[Z_{\xi}], \tag{6.4}$$

where the prime on the function denotes a derivative with respect to its argument.

The two expressions for the fractional derivative of the function can be expressed as phase space integrals

$$\begin{aligned}\partial_t^\alpha[F] &= \lim_{\Delta t\to 0}\int dzF(z)\frac{[\rho_\xi(z,t+\Delta t\mid z_0)-\rho_\xi(z,t\mid z_0)]}{\Delta t^\alpha}\\ &= \int dzF(z)\partial_t^\alpha[\rho_\xi(z,t\mid z_0)]\end{aligned} \tag{6.5}$$

and

$$F'(Z_\xi)\partial_t^\alpha[Z_\xi] = \int dzF'(z)\partial_t^\alpha[z]\rho_\xi(z,t\mid z_0).$$

This last expression can be integrated by parts and setting the boundary terms to zero, as we did earlier, gives us

$$F'(Z_\xi)\partial_t^\alpha[Z_\xi] = -\int dzF(z)\frac{\partial}{\partial z}[\partial_t^\alpha[z]\rho_\xi(z,t\mid z_0)]. \tag{6.6}$$

Again, since the analytic function $F(z)$ is arbitrary, comparing equations (6.5) and (6.6) we can equate coefficients within the integrand to obtain the fractional continuity equation

$$\partial_t^\alpha[\rho_\xi(z,t\mid z_0)] + \frac{\partial}{\partial z}[\partial_t^\alpha[z]\rho_\xi(z,t\mid z_0)] = 0. \tag{6.7}$$

This is a generalization of the continuity equation, which measures the flux of 'stuff' into and out of a region of space at a given time. The generalized continuity equation takes into account memory effects in the generalized flux through a local region of space. The dynamics producing that generalized flux is obtained by substituting from the FSDE into equation (6.7) to obtain

$$\partial_t^\alpha[\rho_\xi(z,t\mid z_0)] + \frac{\partial}{\partial z}[\{G(z)+g(z)\xi(t)\}\rho_\xi(z,t\mid z_0)] = 0. \tag{6.8}$$

Following the procedure implemented in Section 4.4, we average equation (6.8) over an ensemble of realizations of the random parameter $\xi(t)$, to obtain

$$\begin{aligned}\partial_t^\alpha[P(z,t\mid z_0)] &= -\frac{\partial}{\partial z}[G(z)P(z,t\mid z_0) + g(z)\langle\xi(t)\rho_\xi(z,t\mid z_0)\rangle_\xi]\\ &= \mathcal{L}_0 P(z,t\mid z_0) + \langle\mathcal{L}_\xi\rho_\xi(z,t\mid z_0)\rangle_\xi,\end{aligned} \tag{6.9}$$

where we have introduced the operators defined earlier. Assuming the fluctuations are given by a Wiener stochastic process, that is, by fluctuations with Gaussian statistics, delta correlated in time, the correlation term on the RHS, using the Stratonovich definition of the integral, reduces to a second-order partial differential. Equation (6.9) then becomes a FFPE:

$$\partial_t^\alpha[P(z,t\mid z_0)] = \mathcal{L}_{\mathrm{FP}} P(z,t\mid z_0), \tag{6.10}$$

with the Fokker–Planck operator $\mathcal{L}_{\mathrm{FP}}$ defined by

$$\mathcal{L}_{\mathrm{FP}} = -\frac{\partial}{\partial z}[G(z)] + D\frac{\partial}{\partial z}\left[g(z)\frac{\partial}{\partial z}g(z)\right]. \tag{6.11}$$

This is called the FFPE because the deterministic part of the FSDE can be expressed in terms of the gradient of a potential function $U(z)$

$$G(z) = -\frac{\partial U(z)}{\partial z}, \tag{6.12}$$

in which case the FSDE is a fractional generalization of the multiplicative Langevin equation.

The details of a mechanical model that can form the microscopic basis of such a generalization is given by Grigolini et al. [133], where the time separation between the macroscopic and microscopic dynamics breaks down, due to the non-integrability of the equations of motion generated by the microscopic Hamiltonian. The time scale of the microscopic chaos diverges to a macroscopic value, thereby resulting in an overlap of the microscopic and macroscopic time scales. This time scale overlap induces a memory into the dynamics, which is captured by the fractional time derivative. Example calculations for the eigenvalue solution to the FFPE are given elsewhere [166, 352, 362] and need not concern us for application of these ideas to allometry.

## 6.2 Fractional kinetic equations

The fully fractional phase space equations (FFPSEs) are more general than the FFPEs, because in addition to the fractional derivative in time, they also have fractional derivatives in the system variables. Such equations have a remarkably different dynamics from those described by a FFPE, because the trajectories generated by the dynamic equations can themselves be fractal, due to their explicit chaotic nature. Therefore, if one follows the approach used to derive the FFPE, the simplifying assumption made in going from equation (6.9) to equation (6.10) breaks down and a new approach must be adopted. Zaslavsky [387, 388], as well as Grigolini et al. [133], recognized that when the dynamic trajectories are fractal the equations for the PDF must be in terms of fractional operators. Zaslavsky was the first to systematically relate averages of the phase space distribution function, over ensembles of fractal (chaotic) trajectories, to determine the evolution of the PDF in terms of fractional derivatives. His review of the area is well worth studying, but for the purposes of allometry it is sufficient to merely sketch his arguments leading to what he called the *fractional kinetic equations* (FKEs).

We sketch the arguments of Zaslaveky and co-workers [279] to obtain the phase space equations for the PDF when the underlying dynamics are chaotic and consequently the trajectories are fractal. We begin with the chain condition of Bachelier, Smoluchowski, Chapman and Kolmogorov [224]:

$$P(z_2,t_2 \mid z_1,t_1) = \int dyP(z_2,t_2 \mid y,t')P(y,t' \mid z_1,t_1), \tag{6.13}$$

where $P(z_2, t_2 \mid z_1, t_1)$ is the probability density that the variable $z$ undergoes a transition from $z_1$ at time $t_1$ to $z_2$ at time $t_2$, through the value $y$ at the intermediate time $t'$. The BSCK chain condition was first introduced by Bachelier [13] in his thesis (under the direction of Poincaré) on stock market speculation. It was later discussed independently and with varying levels of mathematical rigor by Smoluchowski [301], Chapman [66] and Kolmogorov [169].

Zaslavsky [387] introduces a wandering process along a fractal space–time set by assuming the existence of the limit

$$\partial_t^\alpha P(z,t) = \lim_{\Delta t \to 0} \frac{1}{\Delta t^\alpha} \{P(z, z_0; t + \Delta t) - P(z, z_0; t)\}; \quad 0 < \alpha \leq 1 \tag{6.14}$$

for the fractional derivative in time, assuming the process is stationary and using the notation

$$P(z, t \mid z_0, t_0) = P(z, z_0; t - t_0). \tag{6.15}$$

The fractal nature of the set is generated by the ensemble of chaotic trajectories originating with the underlying non-integrable Hamiltonian. Inserting the $\Delta t$-limit into the BSCK chain condition enables us to relate the fractional time derivative to the limit over the space integrals

$$\partial_t^\alpha P(z,t) = \lim_{\Delta t \to 0} \frac{1}{\Delta t^\alpha} \left\{ \int dy \left[ P(z, y; \Delta t) - \delta(z - y) \right] P(y; t) \right\}, \tag{6.16}$$

which can be simplified using the generalized Taylor expansion of the transition probability in an infinitesimal interval $|z - y|$:

$$P(z, y; \Delta t) = \delta(z - y) + A(y; \Delta t)\delta^{(\beta)}(z - y) + B(y; \Delta t)\delta^{(\beta+1)}(z - y) \tag{6.17}$$

for an appropriate fractal dimension characterized by $0 < \beta \leq 1$.

We introduce the functions resulting from the fractional delta functions in equation (6.17) being inserted into the BSCK chain condition

$$\mathcal{A}(z) \equiv \lim_{\Delta t \to 0} \frac{A(z; \Delta t)}{\Delta t^\alpha} = \lim_{\Delta t \to 0} \int dy \frac{|z - y|^\beta}{\Delta t^\alpha} P(z \mid y; \Delta t), \tag{6.18}$$

and

$$\mathcal{B}(z) \equiv \lim_{\Delta t \to 0} \frac{B(z; \Delta t)}{\Delta t^\alpha} = \lim_{\Delta t \to 0} \int dy \frac{|z - y|^{\beta+1}}{\Delta t^\alpha} P(z \mid y; \Delta t). \tag{6.19}$$

These two functions characterize the local properties of the meandering process in the limit $\Delta t \to 0$. Zaslavsky [387, 388] explains that the limit appears as a result of the fractal dimensionality of the space–time set along which the particle (state of the system) is wandering and we refer the reader to his papers for a more detailed discussion of nonlinear dynamics.

We bypass much of the mathematical detail and note that by inserting equations (6.18) and (6.19) into (6.16) Zaslavsky was able to obtain

$$\lim_{\Delta t\to 0}\frac{1}{\Delta t^{\alpha}}\int dyA(y;\Delta t)\delta^{(\beta)}(z-y)P(y;t)=\int dy\mathcal{A}(y)\delta^{(\beta)}(z-y)P(y;t)$$
$$\Rightarrow \frac{d^{\beta}}{d(-z)^{\beta}}[\mathcal{A}(z)P(z;t)]$$

and

$$\lim_{\Delta t\to 0}\frac{1}{\Delta t^{\alpha}}\left\{\int dyB(y;\Delta t)\delta^{(\beta+1)}(z-y)P(y;t)\right\}=\int dy\mathcal{B}(y)\delta^{(\beta+1)}(z-y)P(y;t)$$
$$\Rightarrow \frac{d^{\beta+1}}{d(-z)^{\beta+1}}[\mathcal{B}(z)P(z;t)]$$

where these fractional derivatives on the RHS are replaced with symmetric Reisz–Feller fractional derivatives, which require a bit more discussion to obtain [388]:

$$\partial^{\beta}_{|z|}=-\frac{1}{2\cos(\pi\beta/2)}\left[\frac{d^{\beta}}{dz^{\beta}}+\frac{d^{\beta}}{d(-z)^{\beta}}\right]. \tag{6.20}$$

The resulting one-dimensional form of the FKE is [279]

$$\partial^{\alpha}_{t}P(z,t)=\partial^{\beta}_{|z|}[\mathcal{A}(z)P(z;t)]+\partial^{\beta+1}_{|z|}[\mathcal{B}(z)P(z;t)], \tag{6.21}$$

with the fractional indices in the domain $0<\alpha,\ \beta\le 1$. Recall that the Reisz–Feller fractional derivative $\partial^{\beta}_{|z|}[f(z)]$ is most readily defined in terms of its Fourier transform $-|k|^{\beta}\tilde{f}(k)$, where $\tilde{f}(k)$ is the Fourier transform of $f(z)$.

### 6.2.1 Time-dependent Lévy process

The argument leading from the FKE for the PDF, in the case where the 'fractional velocity' is zero and the strength of the 'fractional diffusion' is constant

$$\mathcal{A}(z)=0 \quad \text{and} \quad \mathcal{B}(z)=K_{\beta}, \tag{6.22}$$

in both the phase space of the dynamic variable and time, to its scaling solution has been presented elsewhere [166, 353]. Given the simplicity of the argument and for the sake of completeness we present it here as well. The FKE is [166, 388]:

$$\partial^{\alpha}_{t}[P(z,t)]=K_{\beta}\partial^{\beta+1}_{|z|}[P(z,t)], \tag{6.23}$$

where $\partial^{\alpha}_{t}[\cdot]$ is taken to be the Caputo fractional derivative in time and $\partial^{\beta+1}_{|z|}[\cdot]$ is the Reisz–Feller fractional derivative in one space dimension. This equation is solved as an initial value problem, where $P_0(z)$ is the initial PDF,

$$P(z,t=0)=P_0(z) \tag{6.24}$$

typically taken to be the delta function $\delta(z)$; and $K_\beta$ is a generalized diffusion coefficient. The probability that the dynamic variable $Z(t)$ lies in the interval $(z, z + dz)$ at time $t$ is the phase space quantity $P(z,t)dz$. It is not necessary to review the fractional calculus in order to understand the solution to equation (6.23). The solution may be obtained using the scaling properties of the FKE.

The Fourier transform of the symmetric Reisz–Feller operator $\partial_{|z|}^{\beta+1}[\cdot]$ acting on an analytic function $f(z)$ is [166, 293]

$$\mathcal{FT}\{\partial_{|z|}^{\beta+1}[f(z)], k\} = -|k|^{\beta+1}\tilde{f}(k), \tag{6.25}$$

where $\tilde{f}(k)$ is the Fourier transform of $f(z)$. The Laplace transform of a Caputo fractional time derivative acting on the analytic function $g(t)$ is

$$\mathcal{LT}\{\partial_t^\alpha[g(t)]; u\} = u^\alpha \hat{g}(u) - u^{\alpha-1}g(0), \tag{6.26}$$

for $0 < \alpha < 1$, where $\hat{g}(u)$ is the Laplace transform of $g(t)$. Consequently, the phase space dynamics given by equation (6.23) can be expressed as

$$u^\alpha P^*(k,u) - u^{\alpha-1}\tilde{P}(k,t=0) = -K_\beta|k|^{\beta+1}P^*(k,u) \tag{6.27}$$

and the asterisk denotes the Fourier–Laplace transform of the PDF. Therefore, the solution to equation (6.23), when $P_0(z) = \delta(z)$, using the unit Fourier transform of the delta function

$$\tilde{P}(k,t=0) = 1,$$

in Fourier–Laplace space is

$$P^*(k,u) = \frac{u^{\alpha-1}}{u^\alpha + K_\beta|k|^{\beta+1}}. \tag{6.28}$$

The PDF that solves the FFPSE is given by the inverse Fourier–Laplace transform of equation (6.28).

We note that the space–time representation of the solution to the FKE for various combinations of $\alpha$ and $\beta$ and potential functions are reviewed by Klafter and Metzler [166], who show how to derive equation (6.23) using the continuous time random walk (CTRW) of Montroll and Weiss [223]. The derivation of the FKE using subordination theory is also available [362].

The inverse Laplace transform of equation (6.28) yields the MLF, just as obtained in Section 5.2:

$$\tilde{P}(k,t) = E_\alpha(-K_\beta|k|^{\beta+1}t^\alpha), \tag{6.29}$$

which is the characteristic function for a ML process. The inverse Fourier transform of the characteristic function yields the PDF

$$P(z,t) = \mathcal{FT}^{-1}\{E_\alpha(-K_\beta|k|^{\beta+1}t^\alpha); z\} \tag{6.30}$$

When $\alpha = 1$ we know that the MLF reduces to an exponential, in which case the solution, given by equation (6.30), is the characteristic function for the stable Lévy distribution in space, with a Lévy index $0 < \beta + 1 \leq 2$ and a "width parameter" that increases linearly with time $K_\beta t$:

$$P(z,t) = \mathcal{FT}^{-1}\{\exp[-K_\beta t|k|^{\beta+1}];z\}. \tag{6.31}$$

The series representation for the Lévy distribution is given in a number of places, e.g. [224].

It is interesting to note that the Lévy PDF appears in a number of complex processes including turbulence [296], dynamic spatial redistribution of individuals [45]. A variety of other solutions to the FKE, for $0 < \alpha < 1$, have been obtained by Mainardi [197], including the inverse Fourier transform for $\beta = 1$, in which case the solution asymptotically relaxes as the inverse power law $t^{-\alpha/2}$.

## 6.3 Allometry scaling solution

We can use the solution of the preceding section, with some care, to determine the behavior of the allometry PDF. The above derivation has been done in one variable for notational convenience, however, for application to the allometry problem we need to extend the FKE to the two variable case. Of course, the stochastic variables of interest here are not the position and momentum of physics; they are, in fact, the measures of functionality and size of the allometry phenomena being investigated.

We anticipate the stochastic behavior of the two allometry variables to be similar, since they are separately related to the underlying system's complexity. Recall that the average size variable was modeled by a nonlinear rate equation and its stochastic generalization was given in Section 4.3. The trajectories for the size variable of individuals are therefore expected to be solutions to nonlinear equations of motion and consequently are generically chaotic. If all the variability in the allometry relation were assumed to reside in the positive definite size variable, it would be reasonable to conclude that the size PDF satisfies a FKE of the form

$$\partial_t^\alpha[P(x;t)] = \frac{d^\beta}{dx^\beta}[\mathcal{A}(x)P(x;t)] + \frac{d^{\beta+1}}{dx^{\beta+1}}[\mathcal{B}(x)P(x;t)], \tag{6.32}$$

following the discussion in the previous section. Note that this fractional derivative is defined by

$$\frac{d^\beta}{dx^\beta}f(x) = \frac{1}{\Gamma(-\beta)}\int_0^x \frac{f(z)}{(x-z)^{\beta+1}}, \tag{6.33}$$

whose Fourier transform is given by [279]

$$\mathcal{FT}\left\{\frac{d^\beta}{dx^\beta}f(x);k\right\} = (-ik)^\beta\tilde{f}(k). \tag{6.34}$$

On the other hand, unlike the size of the individual, there is no reason to assume individual functionality has a fractal trajectory. Therefore we might assume the functionality PDF to be described by a FFPE:

$$\partial_t^\alpha [P(y;t)] = \partial_y^\mu [A(y)P(y;t)] + \partial_y^{\mu+1} [B(y)P(y;t)], \tag{6.35}$$

with the fractional derivative index for the functionality in the interval $0 \le \mu \le 1$. But neither equation (6.32) nor (6.35) tells the entire story, because the system's size and functionality are tied together through their underlying dependence on the system's complexity. Consequently, the challenge is to rigorously construct the phase space equation for the joint PDF $P(x,y;t)$.

In order to construct the two variable phase space equation each of the above coefficients must be generalized to reflect the possible coupling between the allometry variables. With this in mind we write

$$\begin{aligned} \partial_t^\alpha [P(x,y;t)] = {} & \frac{d^\beta}{dx^\beta} [\mathcal{A}(x,y)P(x,y;t)] + \frac{d^{\beta+1}}{dx^{\beta+1}} [\mathcal{B}(x,y)P(x,y;t)] \\ & + \partial_y^\mu [A(x,y)P(x,y;t)] + \partial_y^{\mu+1} [B(x,y)P(x,y;t)] \end{aligned} \tag{6.36}$$

with no more justification than an inspection of equations (6.32) and (6.35), and the requirement that $\mathcal{A}(x,y) \to \mathcal{A}(x)$ and $\mathcal{B}(x,y) \to \mathcal{B}(x)$ by integrating over $y$, as well as, $A(x,y) \to A(y)$ and $B(x,y) \to B(y)$ by integrating over $x$. Equation (6.36) is a conjecture and the actual two-variable form is left as an open challenge to construct. Consequently, the results contained in the remainder of this section can be reasonably regarded as reasoned speculations, which hopefully the reader will find to be of some value.

We expect, due to the fact that the allometry measures are all determined when the system of interest is mature, that the FKE of interest is in the asymptotic ($t \to \infty$) regime. Consequently, the equation for the steady-state solution to the FKE is determined by

$$\lim_{t\to\infty} \partial_t^\alpha [P(x,y;t)] = 0 \quad \Rightarrow \quad P_{ss}(x,y) = \lim_{t\to\infty} P(x,y;t), \tag{6.37}$$

A rough interpretation of the coefficients in the FKE is obtained in parallel to the traditional derivation of the FPE where $\mu = \beta = 1$, in which case the $A,\mathcal{A}$-terms give the average fractional velocity and the $B,\mathcal{B}$-terms yield fractional diffusion from the average strength of a random force. The average fractional velocity term for the size is zero, since growth has abated for a mature individual, and what dynamics remain are in the chaos-induced fractional diffusion, so we assume

$$\mathcal{A}(x,y) = 0 \quad \text{and} \quad \mathcal{B}(x,y) = D_x. \tag{6.38}$$

By the same token, the deterministic part of the functionality has also reached equilibrium so that the functionality in the steady state is determined by random variability

$$A(x,y) = 0 \quad \text{and} \quad B(x,y) = D_y. \tag{6.39}$$

Consequently, we obtain for the allometry FKE the steady-state equation

$$\partial_y^{\mu+1}[P_{ss}(x,y)] = -K\frac{d^{\beta+1}}{dx^{\beta+1}}[P_{ss}(x,y)]; \quad K \equiv \frac{D_x}{D_y}, \tag{6.40}$$

which is of the form equation (6.23), with time in the former equation replaced with functionality and $\alpha = \mu + 1$ in the latter, in which case this equation may be solved using a similar methodology.

The double transformed solution to the allometry FKE, given by a suitably interpreted equation (6.28), is provided by directly utilizing the scaling properties of equation (6.29). This is done by considering the formal solution to be the real part of the inverse Fourier transform of the MLF for the allometry FKE:

$$P_{ss}(x,y) = \mathrm{Re}[\mathcal{FT}^{-1}\{E_{\mu+1}(-y^{\mu+1}Kk^{\beta+1}e^{i(\beta+1)\pi});x\}]. \tag{6.41}$$

The series expansion for the MLF, when inserted into equation (6.41), allows us to write for the scaled form of the PDF

$$P_{ss}(Ax,By) = \mathrm{Re}\left[\sum_{n=0}^{\infty}\frac{(Ke^{i\beta\pi}By^{\mu+1})^n}{\Gamma(n[\mu+1]+1)}\cdot\frac{\Gamma(n[\beta+1]+1)}{|Ax|^{n(\beta+1)+1}}\right], \tag{6.42}$$

where the second factor in the summation is the result of applying a Tauberian theorem to the inverse Fourier transform of $|k|^{n(\beta+1)}$. A scaling equation emerges from equation (6.42), when the parameters satisfy the equality $A = B^{1/(\beta+1)}$, simplifying to

$$P_{ss}(B^{1/(\beta+1)}x, By) = \frac{1}{B^{1/(\beta+1)}}P_{ss}(x,y). \tag{6.43}$$

If we now select the $B$ parameter to be the inverse power of the functionality, $B = 1/y^{(\mu+1)}$, we can rearrange terms to write

$$P_{ss}(x,y) = \frac{1}{y^{\frac{\mu+1}{\beta+1}}}P_{ss}\left(\frac{x}{y^{\frac{\mu+1}{\beta+1}}},1\right). \tag{6.44}$$

Finally, the PDF that solves the allometry FKE in terms of the similarity variable $x/y^{\gamma}$, satisfies the scaling equation

$$P_{ss}(x,y) = \frac{1}{y^{\gamma}}F\left(\frac{x}{y^{\gamma}}\right) \quad \text{with } \gamma = \frac{\mu+1}{\beta+1}, \tag{6.45}$$

where we have simplified the notation by introducing the empirical function

$$F\left(\frac{x}{y^{\gamma}}\right) \equiv P_{ss}\left(\frac{x}{y^{\gamma}},1\right). \tag{6.46}$$

The function $F(\cdot)$ in equation (6.45) is left unspecified, but it is assumed to be analytic in the similarity variable $x/y^{\gamma}$. As stated in Chapter 4, a standard diffusion process

$Z(t)$ is the displacement of a diffusing particle from its initial position at time $t$, and for vanishing small dissipation the scaling parameter is $\gamma = 1/2$ and the functional form of $F(\cdot)$ is a Gauss distribution. However, for general complex phenomenon there is a broad class of distributions for which the functional form of $F(\cdot)$ is not Gaussian and the scaling index $\gamma \neq 1/2$. For example, a stable Lévy process. [224, 276, 293, 391] scales in this way and the Lévy index is in the range $[0, 2]$, with the equality holding for the Gauss distribution and the scaling index is related to the Lévy index by $1/\gamma$.

It is noteworthy that the scaling parameter $\gamma = \frac{\mu+1}{\beta+1}$ in the steady-state PDF gives the relative contribution of the fluctuations in functionality to those in the size. This balance can be interpreted as a system having two interacting networks that characterize its complex dynamic behavior, both of which are stochastic. However, the magnitude of the scaling exponent determines which of the two is the more random in the steady state.

### 6.3.1 What about allometry relations?

Banavar et al. [19] used scaling theory in which the size $x$ is interpreted as the TBM, population abundance, or both and functionality $y$ is the region of area over which the population roams. They obtain an expression for the PDF similar in form to equation (6.40). A sequence of four additional hypotheses were made to establish a framework for the analysis of diverse empirical macro-ecological laws. The fractional calculus approach leading to equation (6.40) is in one sense less ambitious than the scaling theory of Banavar et al. [19], but has the virtue of being a tool by which to systematically investigate the influence of the environment on the process of interest.

We write the steady-state solution to the allometry FKE obtained in the previous section:

$$P_{ss}(x, y) = \frac{1}{y^{\gamma}} F\left(\frac{x}{y^{\gamma}}\right). \tag{6.47}$$

The asymptotic average size for a given functionality is obtained using the steady-state PDF

$$\langle X \rangle = \int x P_{ss}(x, y) dx = y^{\gamma} \overline{X},$$

which can be inverted to yield

$$y = a \langle X \rangle^{b}, \tag{6.48}$$

in agreement with the empirical AR given by equation (1.19), if $y$ is interpreted as the average functionality $\langle Y \rangle$. The theoretical allometry coefficient in equation (6.48) is given in terms of the average of the first moment with respect to the unknown PDF over the similarity variable $q = x/y^{\gamma}$:

$$\overline{X} = \int qF(q)dq, \tag{6.49}$$

so that

$$a = 1/\overline{X}^{b} \tag{6.50}$$

and the theoretical allometry exponent is

$$b = 1/\gamma = \frac{\beta + 1}{\mu + 1}. \tag{6.51}$$

Therefore, the scaling properties of the PDF solution to the steady-state allometry FKE entails the allometry relations.

The statistical variability in the system variable increases with increasing fractional index. Consequently the scaling index of the steady-state PDF provides a direct measure of the relative variability of the system's functionality and size. Most processes from natural history have $b < 1$, and consequently, the FKE interpretation of the AR would have the strength of the fluctuations in the functionality greater than those in size. For example, the variability in average metabolic rate of mammals, or the wing span of birds, is greater than their variations in size. Moreover, the variation in the size of an organ, or predator population, in the AR, is greater than the variability in the size of the host organism in the first case and of the prey population in the second.

On the other hand, for the urban ARs, where $b > 1$, the strength of the fluctuations in city size exceeds those in the functionality measures of the city, such as wages and innovation. One might draw the inference that the larger cities stabilize the variability in wages, innovation and other measures of urban functionality.

The fact that the empirical allometry exponents cluster around certain values for given functionalities, such as basal metabolic rates, strongly suggests that the existence of a narrow range of statistical variation ratios necessary to retain stability. For the metabolic rate AR it is clear that more random fluctuations in the metabolic rate are tolerable than in the size of the animal. This may well have to do with the tolerance of fractal processes to errors in the underlying process, which we discussed in Chapter 1.

Here we note that the allometry coefficient is determined by equation (6.50), the average similarity variable. We emphasize that the similarity variable is precisely the quantity defined as the SAMI measure of urban variability, discussed in Section 3.2.6. The present statistical analysis explains why a newly devised data processing technique, simultaneously developed to describe urban variability [36] by one group and species variability [351] by another, is so successful in both. What is common to the mechanistically different phenomena is the scaling behavior of the underlying statistics.

## 6.4 Entropy entails allometry

A FPSE for anomalous diffusion, containing the influence of long-term memory on the dynamics of the PDF, was first discussed in 1982 by West and Seshadri [336] in a physical context. The resulting FDE [166, 345] has subsequently found a variety of uses in the physical sciences and elsewhere. The PDF that solves a class of FKEs, satisfies the scaling equation (6.47). Note that this scaling of the PDF is also a realization of the RGT relation for the random variable

$$Z(\lambda t) = \lambda^{\gamma} Z(t), \tag{6.52}$$

where the scaling is interpreted in terms of the PDF and not the dynamic variable.

Entropy is often treated as a mysterious quantity, regardless of the number of books written to demystify it, both inside [257, 258] and outside [47] statistical physics. The introduction of entropy shifts perspective from the energy-based idea of a mechanical state, with a single trajectory, to the statistical notion of a "state" involving an ensemble of trajectories. Thus, entropy is intimately related to the idea of probability and characterizes the system as a whole in a variety of possible futures. As early as 1910 Klein [168] summarized entropy dynamics in a number of ways, including: transforming a system from a state of variability to one of uniformity; as an irreversible change from a more probable state to one that is less probable, with the direction of time being thereby determined; as the passage of a system from a state somewhat regulated to one less regulated; as a decrease in the system's order. Consequently, the differential calculus that is used to describe the dynamics of mechanical systems is replaced by the probability calculus to describe complex, but not necessarily mechanical, systems.

All naturally occurring phenomena are irreversible, as determined by entropy and the second law of thermodynamics. Nothing occurs in nature that is reversible, which, because the dynamic equations of Newton are reversible, has resulted in over a century of discussion, concerning how to make the mechanical world view consistent with observation, that is, to account for the fact that time is unidirectional. More formally, the issue is how can physics provide a statistical mechanical foundation for entropy? The argument has not as yet been resolved and we leave it there, but for our purposes we note that as a consequence of their complexity and their dependence on averages, allometric phenomena are irreversible.

The Gibbs entropy (Shannon/Wiener information) [42, 290] can be defined in terms of an allometry system's steady-state PDF

$$S(y) = -\int P_{ss}(x,y) \ln P_{ss}(x,y) dx, \tag{6.53}$$

where we have returned to equation (6.45) for this discussion. Inserting the steady-state PDF into the equation for the entropy yields

$$S(y) = S^0 + \gamma \ln y. \tag{6.54}$$

Here we have defined an entropy reference state

$$S^0 \equiv -\int F(q)\ln F(q)dq, \tag{6.55}$$

integrated with respect to the similarity variable $q = x/y^y$. The entropy in equation (6.54) is seen to increase logarithmically in the functionality $y$. Notice the similarity of the entropy in equation (6.54) with that of the Boltzmann equation

$$S = k\ln\Omega + \text{constant}, \tag{6.56}$$

where $\Omega$ is the number of microstates available to the system. Comparing the two expressions for the entropy, we see that the number of microstates available to the system, grows as a power law in the average system size variable, that is, $\Omega = \langle X\rangle^b$, where Boltzmann's constant $k$ is adjusted to make the *bit* the unit of information.

Klein [168] observed that entropy is both the criterion for, and measure of, irreversibility. Moreover, because irreversibility is dependent on statistical properties it must be essentially dependent on average values. Comparing equations (6.48) and (6.54), enables us to write for the change in entropy, relative to the reference state

$$\Delta S(y) \equiv S(y) - S^0 = \ln[\langle X\rangle/\overline{X}]. \tag{6.57}$$

Consequently, the change in the average value of an individual's size at a given functionality is determined by the change in entropy from its reference state at that functionality

$$\langle X\rangle = \overline{X}e^{\Delta S(y)}. \tag{6.58}$$

This is a completely general result that depends only on the scaling behavior of the steady-state PDF. It is also consistent with Klein's qualitative observation, regarding the relation between entropy and average values.

### 6.4.1 Entropy and intraspecies allometry

Huxley [152] assumed the differential growth equations for an organ and the organism of which it is a part, have the same form, with proportional growth rates, for example, a deer's antlers and its TBM; a crab's claws and its TBM, among many others. West and West [349, 351] adapted this assumption and conjectured that the two interdependent networks, size of organ and the TBM of the host network (organism) share the same class of PDFs. The dynamic quantity $Z(t)$, where $t$ is now explicitly the time, can therefore be identified with either of the interdependent networks' dynamic variables.

If we assume that two interdependent networks within an organism share a PDF form given by

$$P(z,t) = \frac{1}{t^{\gamma_z}}F_z\left(\frac{z}{t^{\gamma_z}}\right), \tag{6.59}$$

then the entropy change in time for each network can be written as

$$\langle Z(t)\rangle = \int zP(z,t)dz = t^{\gamma_z}\int qF_z(q)dq = t^{\gamma_z}\overline{\overline{Z}}.$$

Following the algebra of the previous section, we express the average of the dynamic variable in terms of the entropy change produced by that variable:

$$\langle Z(t)\rangle = \overline{\overline{Z}}e^{\Delta S_z(t)}. \tag{6.60}$$

Consequently, both the average functionality and average size change in time with the associated change in entropy. We obtain the entropy change

$$\Delta S_z = \gamma_z \ln t, \tag{6.61}$$

for both the system's functionality and size variables. It is clear that the rate of entropy (information) generation determined by equation (6.61)

$$\lim_{t\to\infty}\frac{d\Delta S_z(t)}{dt} = \lim_{t\to\infty}\frac{\gamma_z}{t} = 0, \tag{6.62}$$

asymptotically vanishes. However the rate at which information generation approaches zero as a function of system size and functionality can be quite different. During the growth process equation (6.60) indicates a sensitive dependence of the average system response to changes in the system complexity. However, it is clear, using equation (6.62), that the time rate of change of the average response also vanishes asymptotically, which is to say, the system asymptotically approaches a steady state. An expression similar to equation (6.60) was obtained by Demetrius et al. [84], where in that case $\Delta S_z(t)$ is related to the evolutionary entropy for both size and functionality.

Using equation (6.61) we obtain the entropy balance relation produced by the interdependence of the dynamic variables

$$\Delta S_x/\gamma_x = \Delta S_y/\gamma_y. \tag{6.63}$$

Note that this relation is a way of representing a self-imposed constraint on the system's complexity, just as done using Taylor's Law in Chapter 3. Inserting the equality from equation (6.63) into the average functionality given by equation (6.60), yields the intraspecies empirical AR

$$\langle Y_i\rangle = \overline{\overline{Y}}e^{\Delta S_y} = \overline{\overline{Y}}e^{\frac{\gamma_y}{\gamma_x}\Delta S_x} = a\langle X_i\rangle^b. \tag{6.64}$$

In obtaining equation (6.64) we have introduced a second theoretical expression for the allometry coefficient

$$a = \overline{\overline{Y}}/\overline{\overline{X}}^b, \tag{6.65}$$

and the theoretical expression for the allometry exponent

$$b = \gamma_y / \gamma_x. \tag{6.66}$$

We have again introduced the index for the single species $i$ in equation (6.64) to emphasize that this is the intraspecies AR. This derivation of the empirical intraspecies AR is solely a consequence of the equality of the scaling properties of the PDFs for the networks of size and functionality.

The rate of entropy generation is determined by the time derivative of equation (6.60) for $X$ and $Y$ and substituting from the entropy balance equation (6.63) yields

$$\frac{1}{\gamma_x \langle X_i \rangle} \frac{d\langle X_i \rangle}{dt} = \frac{1}{\gamma_y \langle Y_i \rangle} \frac{d\langle Y_i \rangle}{dt}. \tag{6.67}$$

Equation (6.67) shows the correspondence of the dynamics, using the probabilistic approach to that of Huxley, with the dynamic variables of organ size and TBM of an individual species member replaced by averages over an ensemble of individuals' organ size and TBM. It is clear from equation (6.66) that $b < 1$ implies that the rate of growth of the average size of an organ is smaller than that of the average size of the host organism.

In the present case the allometry exponent is now the ratio of the power-law indices in the PDFs, which from equation (6.67) can also be interpreted as the ratio of growth rates. Moreover, the allometry coefficient is determined by the scaling functions in the PDFs. The intraspecies AR captures the average response or adaptation of one network (functionality) to the average behavior of another network (size), through the balance of the entropy generated by each of the two interdependent networks. Moreover, it is the scaling in the PDF and not necessarily the geometric scaling of fractal networks that is the origin of the physiologic ARs. Underlying this derivation is the assumption made earlier that this is an information-dominated network and is consequently driven by information (entropy) gradients and not the energy gradient necessary in the fractal geometry derivation of nutrient transport.

## 6.5 Discussion and conclusions

The fractional calculus offers a different way to think about the origin of allometry. The dynamics of allometry phenomena can be treated as interdependent dynamical statistical allometry variables, taking into account the system's complexity by using the fractional calculus. The representation of allometry phenomena in terms of a joint PDF of the system size and functionality has been shown to be very useful. The fully fractional two-variable phase space equation generalizes Zaslavsky's construction of the FKE, but requires a more careful analysis in the future. However, from the FFPSE form, hypothesized in the present chapter, it is clear that the steady-state solution, obtained when the variability in size and functionality are restrictively produced by random forces, have the general scaling behavior shown in equation (6.45). The scaling property of the exact PDF solution to the FKE was shown to entail the empirical interspecies AR, between the average functionality and the average measure of the network size.

The argument for the intraspecies AR is constructed in a related way. It is assumed that the two interdependent networks, size and functionality, are members of the same class of statistical fluctuations. Moreover, that the PDFs in this class of distributions have the scaling form given by equation (6.59). An entropy balance argument resulting from the information exchange of the two networks was shown to give rise to the intraspecies AR. Consequently, one may cautiously assert that the origins of both interspecies and intraspecies ARs lie with the common scaling behavior of the underlying statistics of both size and functionality.

The universality of the origin of the empirical ARs is tied to the scaling of the statistics; but this entailment has not yet been incontrovertibly established. However, the independent derivation of the empirical AR, using the relation between the change in entropy generated by a network and the average variable associated with that network, suggest that the result may, in fact, be of wide spread value.

# Epilogue

This essay is a continuous, if intermittent, argument in support of the allometry/information (A/I) hypothesis, which is contingent on generalizing the notion of force to non-mechanical phenomena. The idea of a social or biological force is very different from that of a mechanical force, although they are often used interchangeably. A mechanical force originates with the concept of an energy gradient, giving rise to the transfer of momentum from a high energy to a low energy configuration. On the other hand, an information (complexity) force is based on the idea of an information (complexity) gradient, giving rise to the transfer of information from a high information to a low information configuration or system. Thus, paraphrasing a conjecture made by the mathematician Norbert Wiener [376] in 1948, a high energy system that is low in information can be controlled by a low energy system that is high in information. Consequently, the A/I hypothesis introduced in Chapter 1 states:

> ...the origin of natural patterns manifest by allometry relations is the imbalance between the complexity associated with a system's functionality and the complexity associated with a system's size.

The heuristic evidence presented in support of the A/I hypothesis consisted of a number of empirical ARs, from a variety of disciplines, including the physical, social and life sciences. The wide variety of exemplars was thought to be necessary to convince the skeptics that no discipline-specific mechanism could realistically account for the existence of the same functional form for the empirical ARs across disciplines. It was, in fact, these discipline-specific mechanisms that generated most of the controversy over whether the allometry parameters had specific constant values. In particular, whether the allometry exponent in physiologic systems has a universal value.

The argument in support of the A/I hypothesis centered on the ill-defined concept of complexity, settling on the working definition of complexity as a balance between regularity and randomness. The stability of the natural patterns arising in allometry was the result of the ARs being expressed in terms of average quantities, where even here the underlying variability persisted. Consequently, the allometry parameters were treated as fluctuating quantities with empirical PDFs. The statistical variability of the allometry exponent around its average value was found to be described by a Laplace distribution for both the metabolic AR and a number of urban ARs. These results suggested the utility of using the probability calculus, generalized to fractional form, to capture the underlying complexity of allometry phenomena. The stability of the ARs could therefore be traced back to the stability of the steady-state PDF solutions to the phase space equations and not to the dynamic stability of an underlying nonlinear differential equation.

The statistical analysis of metabolic data show that empirical ARs exist across multiple species and consequently the form of the ARs are not solely dependent on

DOI 10.1515/9783110535136-007

species-specific mechanisms. Moreover, we established that there are three reasons for the non-existence of a universal value for the allometry exponent. First there is the variability in the values of the allometry exponent obtained from data analysis in Chapter 3. We could rehash the arguments here, but an objective observer would conclude that although there are indications that a particular value of the allometry exponent might be strongly indicated in a specific context, the evidence for universality is not compelling. Even in the case of metabolic ARs one could reasonably make a case of $b = 2/3$ for small animals, $b = 3/4$ for large animal and $b = 1$ for plants, but no one value of $b$ spans the entire range of animal and plant sizes.

Second, the only theories that predict a universal value of the allometry exponent do so to explain the intraspecies AR, not the interspecies AR. There is no first principles theory that derives the empirical AR between the averaged observables. The phenomenological theory presented in Chapter 4 and Chapter 5 successfully derives the interspecies AR and treats the allometry coefficient and allometry exponent as statistical parameters, determined by the scaling PDFs.

Third and last, the allometry exponent and coefficient are independently determined to co-vary using theory [351] and statistical data analysis [118].

In addition, let us not be too critical of phenomenological theories. Recall that thermodynamics is a phenomenological theory that has enjoyed remarkable success in explaining a plethora of complex physical phenomena. Even so, the physics community has not yet been able to derive thermodynamics from the more reductionist (fundamental) statistical mechanics, so that a mechanistic understanding of thermodynamics remains controversial. By the same token, the phenomenological theory of AR presented herein takes the discussion of AR out of the domain of the reductionist approaches, previously used to derive inter- and intra-species metabolic ARs and refocuses it on the statistical properties of the entire system. These analyses indicate a new avenue for the study of physiologic ARs, one that systematically includes both deterministic and statistical aspects, using the probability calculus. This new perspective indicates that the origins of empirical ARs reside in the scaling properties of the PDFs for interdependent complex networks.

In simple physical phenomena the dynamics of a network coupled to the environment is described by a Langevin equation, which is a stochastic differential equation. This mechanical description of the dynamics is equivalent to the phase space dynamics of the PDF, such as given by the FPE. As the network becomes more complex, its past history becomes, more not less, influential, until finally the FPE must be replaced with a FFPE or FKE, whose dynamics are the domain of the fractional calculus. As we saw in Chapter 6 the general solution to a class of FKE's is a scaling PDF, implying that empirical ARs result from scaling in the statistical dynamics of the phenomenon. The influence of the growth history of an organism on the growth of an organ, leads to intraspecies ARs; the history of the size of an organism's influence on the functionality of the organism across species leads to interspecies ARs.

The case laid out in Chapter 6 implies that empirical ARs are a consequence of information transfer between complex information-dominated networks. Of course, we have not rigorously proven that the information increase in one network and information decrease in another network, with which it interacts, as measured by the change in entropy, have the relations obtained. These relations need to be rigorously derived or experimentally verified to replace the intuition on which they are now based. However, the empirical relations are consistent with the PCM in which the transfer of information, between two complex information-dominated networks, have been shown [10, 347] to proceed from the network with the greater information content to the network with the lesser information content.

The phenomenological theory presented herein, provides a foundation for empirical ARs, and incorporates both stochastic and deterministic mechanisms. From this approach we conclude that the empirical AR for the basal metabolic rate, the lifetime of an organism, and the ARs in the myriad of other complex phenomena are not explainable solely in terms of reductionistic mechanisms. In particular the allometry parameters are determined not to be universal. This was supported by the demonstration that the allometry exponent and coefficient co-vary in metabolic systems.

On the other hand, the laws from the theory of probability can have universal forms without entailing universal parameter values; the LFE and the Law of Large Numbers are two examples of universal forms. The empirical AR is, in part, a consequence of the generic statistical behavior of complex networks that depend on the infinitely divisible nature of the PDF [122, 123]. The theory of the empirical ARs presented herein is an application of the fractional probability calculus to the understanding of the origins of the empirical relations between averages of interacting networks making up a complex host network. Consequently, empirical ARs are manifestations of an underlying probabilistic law, whose origin can be traced back to the fractional probability calculus and the balancing of deterministic and stochastic mechanisms. The form of the empirical ARs are not and cannot be solely dependent on reductionist mechanisms.

Consequently, if people want to make more money, participate in the arts and contribute to an expanding intellectual landscape, the urban ARs tell us that they ought to move to a larger city. Of course, the urban ARs also indicate that the price is an accelerated pace of life, greater exposure to epidemics and living with a greater crime rate, but then there is no such thing as a free lunch.

# Bibliography


[1] Adrian R., "Research concerning the probabilities of the errors which happen in making observations, etc.", *The Analyst; or Mathematical Museum* **1**, 93–109 (1809).

[2] Agutter P. S. and J. A. Tuszynski, "Analytic theories of allometric scaling", *The J. Exp. Biol.* **214**, 1055–1062 (2011).

[3] Alberch P., S. J.Gould, G. F. Oster and D. B. Wake, "Size and shape in ontogeny and phylogeny", *Paleobiology* **5**, 296–317 (1979).

[4] Albert R. and A.-L. Barabasi, "Statistical mechanics of complex networks", *Rev. Mod. Phys.* **74**, 48 (2002).

[5] AL-dabaan B., "Scaling laws in biology", http://www.math-physics-tutor.com/web_documents/bader2.pdf.

[6] Allegrini P., P. Paadissi, D. Menicci and A. Gemignani, "Fractal complexity in spontaneous EEG metastable-state transitions: new vistas on integrated neural dynamics", *Front. Physiol.* **1**, 128 (2010). doi: 10.3389/fphys.2010.00128.

[7] Andresen B., J. S. Shiner and D. E. Uehlinger, "Allometric scaling and maximum efficiency in physiological eigen time", *PNAS* **99**, 5822–5824 (2002).

[8] Anderson P. W., "More id different", *Science* **177**, 393 (1972).

[9] Apol M. E. F., R. S. Etienne and H. Olff, "Revisiting the evolutionary origin of allometric metabolic scaling in biology", *Funct. Ecol.* **22**, 1070–1080 (2008).

[10] Aquino G., M. Bolgona, P. Grigolini and B. J. West, "Beyond the death of linear response: $1/f$ optimal information transport", *Phys. Rev. Lett.* **105**, 069901 (2010).

[11] Aquino G., M. Bologna, B. J. West and P. Grigolini, "Transmission of information between complex systems: 1/f resonance", *Phys. Rev. E* **83**, 051130-1 (2011).

[12] Arnold V. I. and A. Avez, *Ergodic Problems of Classical Mechanics*, W. A. Benjamin, New York (1968).

[13] Bachelier L., *Annals scientifiques de l'ecole normale superieure*, sup. (3) No. 1017 (1900); Eng. trans. by A. J. Boness in [73].

[14] Bak P. and S. Boettcher, "Self-organized criticality and punctuated equilibrium", *Physica D* **107**, 143–150 (1997).

[15] Bak P., *How Nature Works*, Oxford University Press (1997).

[16] Ballard F. J., R. W. Hanson and D. S. Kronfeld, "Gluconeogenesis and lipogenesis in tissue from ruminant and nonruminant animals", *Fed. Proc.* **28**, 218–231 (1969).

[17] Banavar J. R., J. Damuth, A. Maritan and A. Rinaldo, "Modeling universality and scaling", *Nature* **420**, 626 (2002).

[18] Banavar J. R., J. Damuth, A. Maritan and A. Rinaldo, "Allometric cascades", *Nature* **421**, 713 (2003).

[19] Banavar J. R., J. Damuth, A. Maritan and A. Rinaldo, "Scaling in ecosystems and the linkage of macroecological laws", *Phys. Rev. Lett.* **98**, 068104 (2007).

[20] Barenblatt G. I., *Scaling Phenomena in Fluid Mechanics*, Cambridge University Press, Cambridge (1994).

[21] Barenblatt G. I. and A. S. Monin, "Similarity principles for the biology of pelagic animals", *Proc. Nat. Acad. Sci. USA* **99**, 10506–10509 (1983).

[22] Barnsley M., *Fractals Everywhere*, Academic Press, Boston (1988).

[23] Barrat S., M. Barthellemy and A. Vespignani, *Dynamical Processes on Complex Networks*, Cambridge University Press, UK (2008).

[24] Bassett D. S., D. L. Greenfield, A. Meyer-Lindenberg, D. R. Weinberger, S. W. Moore and E. T. Bullmore, "Efficient physical embedding of topologically complex information


DOI 10.1515/9783110535136-008

processing networks in brains and computer circuits", *PLoS Comp. Biol.* **6**, 1–14 (2010).
[25] Bassingthwaighte J. B., L. S. Liebovitch and B. J. West, *Fractal Physiology*, Oxford University Press, New York (1994).
[26] Battiston S., M. Puliga, R. Kaushik, P. Tasca and G. Caldarelli, "Debtrank: Too central to fail? Financial networks, the fed and systemic risk", *Sci. Repts.* **2**, 541 (2012).
[27] Batty M. and P. Longley, *Fractal Cities*, Academic Press, London (1994).
[28] Batty M., R. Carvalho, A. Hudson-Smith, R. Milton, D. Smith and P. Steadman, "Geometric scaling and allometry in large cities", in *Proceedings of the 6th International Space Syntax Symposium*, Istanbul, Turkey (2007).
[29] Batty M., "A theory of city size", *Science* **340**, 1418–1419 (2013).
[30] Beggs J. and I. Plenz, "Neuronal avalanches in neocortical circuits", *J. Neurosci.* **23**, 11167–11177 (2003).
[31] Beiu V. and W. Ibrahim, "Does the brain really outperform Rent's rule?", *Proc. IEEE ISCAS*, doi: 10.1109/ISCAS.2008.4541499 (2008).
[32] Belgrano A., A. P. Allen, B. J. Enquist and J. F. Gillooly, "Allometric scaling of maximum population density: a common rule for marine phytoplankton and terrestrial plants", *Ecol. Lett.* **5**, 611–613 (2002).
[33] Bénichou O., C. Loverdo, M. Moreau and R. Voituriez, "Intermittent search strategies", *Rev. Mod. Phys.* **83**, 81 (2011).
[34] Beran J., *Statistics of Long-Memory Processes*, Monographs on Statistics and Applied Probability **61**, Chapman & Hall, New York (1994).
[35] Bettencourt L. M. A., J. Lobo, D. Helbing, C. Kuhnert and G. B. West, "Growth, innovation, scaling and the pace of life in cities", *PNAS* **104**, 7301 (2007).
[36] Bettencourt L. M. A., J. Lobo, D. Strumsky and G. B. West, "Urban scaling and its deviations: revealling the structure of wealth, innovation and crime across cities", *Plos one* **5**, e1354-1 (2010).
[37] Bettencourt L. M. A., "The origins of scaling in cities", *Science* **340**, 1438 (2013).
[38] Blackstone N. W., "Size and time", *Syst. Zool.* **36**, 211–215 (1987).
[39] Bokma F., "Evidence against universal metabolic allometry", *Func. Eco.* **18**, 184–187 (2004).
[40] Boltzmann L., *Wissenschaftliche Abhandlungen*, Hasenohrl Fr. (Ed.), three volumes, Leipzig (1909).
[41] Boxenbaum H., "Interspecies scaling, allometry, physiological time, and the ground plan of pharmacokinetics", *J. Pharmacokin. Biopharm.* **10**, 201–227 (1982).
[42] Brillouin L., *Science and Information Theory*, Academic Press, New York (1962).
[43] Brody S., *Bioenergetics and Growth*, Reinhold, New York (1945).
[44] Brockett R., *Stochastic Control*, Lecture Notes, Harvard University, Cambridge, MA (2009).
[45] Brockmann D., L. Hufnagel and T. Geisel, "The scalling laws of human travel", *Nature* **439**, 462–465 (2006).
[46] Brokmann X., J. P. Hermiere, G. Messin, P. Desbiolles, J. P. Bouchaud and M. Dahan, "Statistical aging and nonergodicity in the fluorescence of single nanocrystals", *Phys. Rev. Lett.* **90**, 120601-1 (2003).
[47] Brooks D. R. and E. O. Wiley, *Evolution as Entropy, Toward a Unified Theory of Biology*, University of Chicago Press, Chicago (1986).
[48] Brown J. H., *Macroecology*, University of Chicago Press, Chicago, IL (1995).
[49] Brown J. H., V. K. Gupta, B. Li, B. T. Milne, C. Restrepo and G. B. West, "The fractal nature of nature: power laws, ecological complexity and biodiversity", *Phil. Trans. R. Soc. Lond. B* **357**, 619–626 (2002).

[50] Brown J. H., J. F. Gillooly, A. P. Allen, V. M. Savage and G. B. West, “Toward a metabolic theory of ecology”, *Ecology* **85**, 1771–1789 (2004).

[51] Brownlee J., “Density of death rate: Farr’s law”, *J. Roy. Soc. Stat. Soc.* **83**, 280–283 (1920).

[52] Brummitt C. D., R. M. D’Souza and E. A. Leicht, “Suppressing cascades of load in interdependent networks”, PNAS **109** (12), E680–689 (2012).

[53] Buchanan M., *Nexus*, W. W. Norton, New York (2002).

[54] Buldyrev S. V., R. Parshani, G. Paul, H. E. Stanley and S. Havlin, “Catastrophic cascade of failures in interdependent networks”, *Nature* **464**, 1025–1028 (2010).

[55] Bush E. C. and J. M. Allman, “The scaling of white matter to gray matter in cerebellum and neocortex”, *Brain Behavior and Evolution* **61**, 1–5 (2003).

[56] Cadenasso M. L., S. T. A. Pickett and J. M. Grove, “Dimensions of ecosystem complexity: heterogeneity, connectivity, and history”, *Ecological Complexity* **3**, 1–12 (2006).

[57] Calder W. W. III, *Size, Function and Life History*, Harvard University Press, Cambridge, MA (1984).

[58] Callender C., “Is time an illusion?”, *Sci. American* **302**, 15–21 (2010).

[59] Cambel A. B., *Applied Chaos Theory*, Academic Press, Boston (1993).

[60] Caputo M., “Distributed order differential equations modeling dielectric induction and diffusion”, *Fract. Calc. Appl. Anal.* **4**, 421 (2001).

[61] Carreras B. A., V. E. Lynch, I. Dobson and D. E. Newman, “Critical points and transitions in an electric power transmission model for cascading failure blackouts”, *Chaos* **12**, 985–994 (2002).

[62] Castelltort S., G. Simpson and A. Darrioulat, “Slope-control aspect of river basins”, *Terra Nova* **21**, 265–270 (2009).

[63] Cavagna A., A. Cimarelli, I. Giardina, G. Parisi, R. Santagati, F. Stefanini and M. Viale, “Scale-free correlations in starling flocks”, *Proc. Natl. Acad. Sci. USA* **107**, 11865 (2010).

[64] Changizi M. A., “Principles underlying mammalian neocortical scalling”, *Biol. Cybern.* **84**, 207–215 (2001).

[65] Chaitin G. J., “On th length of programs for computing finite binary sequences: a statistical consideration”, *J. Assoc. Comp. Mach.* **16**, 145–159 (1969).

[66] Chapman S., “On the law of distribution of molecular velocities, and on the theory of viscosity and thermal conduction, in a non-uniform simple monatomic gas”, *Phil. Trans. Roy. Soc. A* **216**, 279 (1916).

[67] Checkland P. B., *Systems Thinking, Systems Practice*, Wiley, New York (1981).

[68] Chialvo D. R., “Emergent complex neural dynamics”, *Nature Phys.* **6**, 744 (2010).

[69] Chvosta P. and P. Reineker, “Non-Markov noise in barrier-fluctuation model”, *J. Phys. A* **30**, L307 (1997).

[70] Christie P. and D. Stroobandt, “The interpretation and application of Rent’s rule”, *IEEE Trans. VLSI Sys.* (Dec. 2000).

[71] Cohen J. E., T. Jonsson and S. R. Carpenter, “Ecological community description using the food web, species abundance, and body size”, *Proc. Nat. Acad. Sci. USA* **100**, 1781–1786 (2003).

[72] Collard M., A. Ruttle, B. Buchanan and M. J. O’Brien, “Population size and cultural evolution in nonindustrial food-producing societies”, *PLoS ONE* **8**, e72628 (2013).

[73] Cootners P., *The Random Character of the Stock Market*, MIT Univ. Press, Cambridge, MA (1954).

[74] Cope E. D., *The Primary Factors of Organic Evolution*, Open Court Publishing Company (1896).

[75] Cowen G. A., D. Pines and D. Metzler (Eds), *Complexity, Metaphors, Models and Reality*, Addison–Wesley, Reading, Mass. (1994).

[76] Cuvier G., *Recherches sur les ossemens fossils*, Paris (1812).

[77] Cyr H., "Individual energy use and he allometry of population density", in *Scaling in Biology*, pp. 267–295, Brown J. H. and G. B. West (Eds), Oxford University Press, New York (2000).
[78] Cyr H. and S. C. Walker, "An Illusion of mechanistic understanding", *Ecology* **85**, 1802–1804 (2004).
[79] Damuth J., "Population density and body size in mammals", *Nature* **290**, 699–700 (1981).
[80] Damuth J., "Home range, home range overlap, and species energy use among herbivorous mammals", *Biol. J. Linn. Soc.* **15**, 185–193 (1981).
[81] Darveau C. A., R. K. Suarez, R. D. Andrews and P. W. Hochachka, "Allometric cascade as a unifying principle of body mass effects on metabolism", *Nature* **417**, 166–170 (2002).
[82] Darveau C. A., R. K. Suarez, R. D. Andrews and P. W. Hochachka, "Darvearu et al. reply", *Nature* **421**, 714 (2003).
[83] Dawson T. H., "Allometric scaling in biology", *Science* **281**, 751a (1998).
[84] Demetrius L., S. Legendre and P. Harrenöes, "Evolutionary entropy: a predictor of body size, metabolic rate and maximal life span", *Bull. Math. Biol.* **71**, 800–818 (2009).
[85] Dewey T. G., *Fractals in Molecular Biophysics*, Oxford University Press, Oxford (1997).
[86] Dewey F. E., D. Rosenthal, D. J. Murphy, V. F. Froelicher and E. A. Ashley, "Does size matter? Clinical applications of scaling cardiac size and function for body size", *Circulation* **117**, 2279–2287 (2008).
[87] Dicke U. and G. Roth, "Neuronal factors determining high intelligence", *Phil. Trans. R. Soc. B* **371**, 0108 (2015). http://dx.doi.org/10.1098/rstb.2015.0180.
[88] Dodds P. S. and D. H. Rothman, "Scaling, universality and geomorphology", *Ann. Rev. Earth Planet. Sci.* **28**, 1–41 (2000).
[89] Dodds P. S., D. H. Rothman and J. S. Weitz, "Re-examination of the "3/4-law" of metabolism", *J. Theor. Biol.* **209**, 9–27 (2001).
[90] Dubois E., "Sur le rapport du poids de l'encephale avec la grandeur du corps chez mammiferes", *Bull. Soc. Anthropol. (Paris)* **8**, 337 (1897).
[91] Duarte C. M., S. Agustin and R. H. Peters, "An upper limit to the abundance of aquatic organisms", *Oecologia* **74**, 272–276 (1987).
[92] Eldredge N., *Time Frames*, Princeton University Press, Princeton, NH (1985).
[93] Eldredge N. and S. J. Gould, "Punctuated equilibria: an alternative to phyletic gradualism", in *Models in Paleobiology*, pp. 82–115, Schopf T. J. M. (Ed.), Freeman, Cooper and Co., San Francisco, CA (1972).
[94] Eloy C., "Leonardo's rule, self-similarity, and wind-induced stresses in trees", *Phys. Rev. Lett.* **107**, 258101-1, -5 (2011).
[95] Enquist B. J., "Universal scaling in tree and vascular plant allometry: towards a general quantitative theory linking plant form and function from cells to ecosystems", *Tree Phys.* **22**, 1045–1064 (2002).
[96] Etienne R. S., M. E. Apol and H. Olff, "Demystifying the West, Brown & Enquist model of the allometry of metabolism", *Funct. Ecol.* **20**, 394–399 (2006).
[97] Falconer K., *Fractal Geometry*, John Wiley, New York (1990).
[98] Farr W., "A report on the mortality of lunatics", *J. Stat. (Lond.)* **9**, 17–33 (1841).
[99] Feder J., *Fractals*, Plenum, New York (1988).
[100] Feldman H. A. and T. A. McMahon, "The 3/4 mass exponent for energy metabolism is not a statistical artifact", *Resp. Physiol* . **52**, 149–163 (1983).
[101] Flood R. L. and E. R. Carson, *Dealing with Complexity*, 2nd Edition, Plenum Press, New York (1993); 1st Edition (1988).
[102] Fraiman D., P. Balenzuela, J. Foss and D. R. Chialvo, "Ising-like dynamics in large-scale functional brain networks", *Phys. Rev. E* **79**, 061992 (2009).

[103] Fronczak A. and P. Fronczak, "Origins of Tayor's power law for fluctuation scaling in complex systems", *Phys. Rev. E* **81**, 066112-1, -13 (2010).

[104] Ferguson-Gow H., S. Sumner, A. F. G. Bourke and K. E. Jones, "Colony size predicts division of labour in attine ants", *Proc. Roy. Soc. B* **281**, 20141411-1, -9 (2014).

[105] Fialkowski M. and A. Bitner, "Universal rules for fragmentation of land by humans", *Landscape Ecol.* **23**, 1013–1022 (2008).

[106] Fitch W. T., "Skull dimensions in relation to body size in nonhuman mammals: the causal bases for acoustic allometry", *Zool.* **103**, 40–58 (2000).

[107] Fraser S. J. and R. Kapral, "Periodic dichotomous-noise-induced transitions and stochastic coherence", *Phys. Rev. A* **45**, 3412 (1992).

[108] Fürth R., *Investigations on the Theory of the Brownian Movement by Albert Einstien, Ph.D.*, (1926), translated by A. D. Cowper, Dover, New York (1956).

[109] Galileo G., This is in the Dialogue of the Second Day in the *Discorsi* of 1638, the work Galileo wrote while under house arrest by the Inquisition. It was translated as *Dialogues Concerning Two New Sciences* by H. Crew and A. De Salvor in 1914 and reprinted by Dover, New York (1952).

[110] Galton F., "The geometric mean in vital and social statistics", *Proc. R. Soc. London* **29**, 365–367 (1879).

[111] Gauss F., *Theoria motus corporum coelestrium*, Hamburg (1809); Dover Eng. Trans., *Theory of Motion of Heavenly Bodies Moving about the Sun in Conic Sections*, New York (1963).

[112] Gayon J., *Am. Zool.* **40**, 748 (2000).

[113] Gibbs W., *Elementary Principles in Statistical Mechanics*, Ox Bow Press, Woodbridge, Conn. (1981), first published in 1901.

[114] Gingerich P. D., "Arithmetic or geometri normality of biological variation: an empiricial test of theory", *J. Theor. Biol.* **204**, 201–221 (2000).

[115] Glansdorf P. and I. Prigogine, *Thermodyanmic Theory of Structure, Stability and Function*, Wiley, New York (1971).

[116] Glass N. R., "Discussion of calculation of power function with special reference to respiratory metabolism in fish", *J. Fish Res. Board Can.* **26**, 2643–2650 (1969).

[117] Glazier D. S., "Beyond the '3/4-power law': variation in the intra- and interspecific scaling of metabolic rate in animals", *Biol. Rev.* **80**, 611–662 (2005).

[118] Glazier D. S., "The 3/4-power law is not universal: Evolution of isometric, ontogenetic metabolic scaling in pelagoic animals", *BioScience* **56**, 325–332 (2006).

[119] Glazier D. S., "A unifying explanation for diverse metabolic scaling in animals and plants", *Biol. Rev.* **85**, 111–138 (2010).

[120] Glöckle W. G. and T. F. Nonnenmacher, "Fractional integral operators and Fox functions in the theory of viscoelasticity" *Macromolecules* **24**, 6426 (1991).

[121] Glöckle W. G. and T. F. Nonnenmacher, "A fractional calculus approach to self-similar protein dynamics", *Biophys. J.* **68**, 46 (1995).

[122] Gnedenko B. V. and A. N. Kolmogorov, *Limit Distributions for Sums of Independent Random Variables*, translated from Russian by K. L. Chung, Addison-Wesley, Reading, MA (1954).

[123] Gnedenko B. V., *The Theory of Probability*, translated by B. D. Seckler, Chelsea Pub., New York (1962).

[124] Gödel K., "Uber formal unintscheibare Satze der Principa Mathemaica und verwandter Systeme I", *Monotshefte fur Mathematik and Physik* **38**, 173–198 (1931).

[125] Goel N. S., S. C. Maitra and E. W. Montroll, "On the Volterra and other nonlinear models of interacting populations", *Rev. Mod. Phys.* **43**, 231–276 (1971).

[126] Goldbeter A., *Biochemical Oscillations and Cellular Rhythms*, Cambridge University Press, New York (1996).

[127] Gotccitas V. and P. Colgan, "Predator foraging success and habitat complexity: quantitative test of the threshold hypothesis", *Oecologia* **80**, 158–166 (1989).

[128] Gould S. J., "Allometry and size in ontogeny and phylogeny", *Biol. Rev. Cam. Philos. Soc.* **41**, 587–640 (1966).

[129] Gould S. J., "Geometric similarity in allometric growth: a contribution to the problem of scaling in the evolution of size", *The Am. Natur.* **105**, 113–136 (1971).

[130] Graham J. H., K. Shumazu, J. E. Emien, D. C. Freeman and J. Merkel, "Growth models and the expected distribution of fluctuating symmetry", *Biol. J. Linn. Soc.* **80**, 57–65 (2003).

[131] Greenewalt C. H., "The flight of birds", *Trans. Am. Phil. Soc.* **94**, 1–67 (1975).

[132] Grenfell, B. T., C. S. Williams, O. N. Bjornstad and J. R. Banavar, "Simplifying biological complexity", *Nature* **21**, 212–213 (2006).

[133] Grigolini P., A. Rocco and B. J. West, "Fractional calculus as a macroscopic manifestation of randomness", *Phys. Rev. E* **59**, 2303 (1999).

[134] Gutenberg B. and C. F. Richter, "Magnitude and energy of earthquakes", *Annali di Geofisica* **9**, 1–15 (1956).

[135] Hack J. T., "Studies of longitudinal profiles in Virginia and Maryland", *US Geol. Surv.Prof. Pap. B* **294**, 52 (1957).

[136] Haase M., C. G. Hubneer, E. Reuther, A. Herrmann, K. Mullen and Th. Basche, "Exponential and power-law kinetics in single-molecule fluorescence intermittency", *J. Phys. Chem. B* **108**, 10445 (2004).

[137] Hatton I. A., K. S. McCann, J. M. Fryxell, T. J. Davies, M. Smerlak, A. R. E. Sinclair and M. Loreau, "The predator-prey power law: biomass scaling across terrestrial and aquatic biomes", *Science* **349**, aac6284-1, -12 (2015).

[138] Hausdorff J. M., P. L. Purdon, C.-K. Peng, Z. Ladin, J. Y. Wei and A. L. Goldberger, "Fractal dynamics of human gait: stability of long-range correlations", *J. Appl. Physiol.* **80**, 1448–1457 (1996).

[139] Hayek L. C. and M. A. Buzas, *Surveying Natural Populations*, Columbia University Press, New York (1997).

[140] He J. H., "An allometric scaling law between gray matter and white matter of cerebral cortex", *Chaos, Solitons and Fractals* **27**, 864–867 (2006).

[141] Hemmingsen A. M., "Energy metabolism as related to body size and respiratory surfaces, and its evolution", *Rep. Steno. Memor. Hosp. Nork. Insul.* **9**, 1–110 (1960).

[142] Hempleman S. C., D. L. Kilgore, C. Colby, R. W. Bavis and F. L. Powell, "Spike firing allometry in avian intrapulmonary chemoreceptors: matching neural code to body size", *J. Exp. Biol.* **208**, 3065–3073 (2005).

[143] Hess W. R., "Das Prinzip des kleinsten Krafrverbrauchs im Dienste Hamodynamischer Forschung", *Archiv Anat. Physiol.* (1913).

[144] Heusner A. A., "Energy metabolism and body size: I. Is the 0.75 mass exponent of Kleiber's equation a statistical artifact?", *Resp. Physiol.* **48**, 1–12 (1982).

[145] Heusner A. A., "Size and power in mammals", *J. Exp. Biol.* **160**, 25–54 (1991).

[146] Hill A. V., "The heat of shortening and the dynamic constants of muscle", *Proc. Roy. Soc. London Ser. B* **126**, 136 (1938).

[147] Hill A. V., "The dimensions of animals and their muscular dynamics", *Sci. Prog.* **38**, 209–230 (1950).

[148] Hofman M. A., "Size and shape of th cerebral cortex in mammals. I. The cortical surface", *Brain Behav. Evol.* **27**, 28–40 (1985).

[149] Horton R. E., "Erosional development of streams and their drainage basins: hydophysical approach to quantitative geomorphology", *Geol. Soc. Amer. Bull.* **56**, 275–370 (1945).

[150] Humphreys N. A. (Ed.), *Vital Statistics: A Memorial Volume of Selections from the Reports and Writings of William Farr*, The Sanitory Institute of Great Britain, London (1885).

[151] Hu T.-M. and W. L. Hayton, "Allometric scaling of xenobiotic clearance: uncertainty versus universality", *AAAPS Pham. Sci.* **3**, 1–14 (2001).

[152] Huxley J. S., *Problems of Relative Growth*, Dial Press, New York (1931).

[153] Itô K., *Mem. Amer. Math. Soc.*, 1 (1951).

[154] Jain R., M. C. Rivera and J. A. Lake, "Horizontal gene transfer among genomes: the complexity hypothesis", *PNAS* **96**, 3801–3806 (1999).

[155] Jerison H. J., "Quantitative analysis of evolution of the brain in mammals", *Science* **133**, 1012–1014 (1961).

[156] Jerison H. J., "Allometry, brain size, cortical surface, and convolutedness", in *Primate Brain Evolution*, pp. 77–84, Armstrong E. and Falk O. (Eds), Plenum, New York (1982).

[157] Jones J. H., "Optimization of the mammalian respiratory system: symmorphosis versus single species adaptation", *Comp. Biochem. Physiol. B* **120**, 125–138 (1998).

[158] Jonsson T., J. E. Cohen and S. R. Carpenter, "Food webs, body size and species abundance in ecological community description", *Adv. Ecol. Res.* **36**, 1–84 (2005).

[159] Joosten J. J., F. Soler-Toscano and H. Zenil, "Program-size versus time complexity", *Int. J. Unconventional Computing.* **7**, 353–387 (2011).

[160] Jumarie G., "Modified Riemann–Liouville derivative and fractional Taylor series of nondifferentiable functions further results", *Computers and Math. with Appl.* **51**, 1367–1376 (2006).

[161] Jung Y., E. Barkai and R. J. Silbey, "Current status of single-molecule spectroscopy: theoretical aspects", *Chem. Phys.* **284**, 181 (2002).

[162] Kadanoff L. P., W. Götze, D. Hamblen, et al., "Static phenomena near critical points: theory and experiment", *Rev. Mod. Phys.* **39**, 395–431 (1967).

[163] Kaitaniemi P., "How to derive biological information from the value of the normalization constant in allometric equations", *PLoS ONE* **3** (4), e1932 (2008).

[164] Katz Y., K. Tunstrom, C. C. Ioannou, C. Huepe, I. D. Couzin, "Inferring the structure and dynamics of interactions in schooling fish", Proc. Natl. Acad. Sci. USA **108** (46), 18720–18725 (2011).

[165] Kerkhoff A. J. and B. J. Enquist, "Multiplicative by nature: why logarithmic transformation is necessary in allometry", *J. Theor. Biol.* **257**, 519–521 (2009).

[166] Klafter J. and R. Metzler, "The random walk's guide to anomalous diffusion: a fractional dynamics approach", *Phys. Rept.* **339**, 1–77 (2000).

[167] Kleiber M., "Body size and metabolism", *Hilgarida* **6**, 315–353 (1932).

[168] Klein F., *Physical Significance of Entropy or of the Second Law*, D. van Nostrand Co., NY (1910).

[169] Kolmogorov A. N., "Über die analytischen Methoden in der Wahrscheinlichkeitsrechnung", *Math. Ann.* **104**, 415 (1931).

[170] Kolmogorov A. N., *Comptes Rendus (Dokl.) Akad. Sci. URSS* **26**, 115 (1941).

[171] Kolmogorov A. N., *Doklady Akademii Nauk* **119**, 754 (1965).

[172] Köpf M., R. Metzler, O. Haferkamp and T. F. Nonnenmacher, in *Fractals in Biology and Medicine*, Vol. II, Losa G. A., D. Merlini, T. F. Nonnenmacher and E. R. Weibel (Eds), Birkhäuser, Basel (1998).

[173] Kopletz P., L. Bartosch and L. Schütz, *Introduction to Functional Renormalization Group*, Lectures in Physics **798**, Springer (2010).

[174] Kozlowski J. and M. Konarzewski, "Is West, Brown and Enquisst's model of allometric scaling mathematically correct and biologically relevant?", *Func. Ecol.* **18**, 283–289 (2004).

[175] Kozlowski J. and M. Konarzewski, "West, Brown and Enquisst's model of allometric scaling again: the same questions remain", *Func. Ecol.* **19**, 739–743 (2005).

[176] Labra F. A., P. A. Marquet and F. Bozinovic, "Scaling metabolic rate fluctuations", *PNAS* **104**, 10900–10903 (2007).
[177] Laplace P. S., *Analytic Theory of Probabilities*, Paris (1810).
[178] Langevin P., *Comptes Rendus Acad. Sci.*, 530, Paris (1908).
[179] Laskin N., "Fractional Poisson process", *Comm. Non. Sci. & Num. Sim.* **8**, 201–213 (2003).
[180] Lasota A. and M. C. Mackey, *Chaos, Fractals and Noise*, Springer-Verlag, New York (1994).
[181] Latka M., B. J. West and W. Jernajczyk, "What a physicist can learn from a somnologist?", *Kosmos* **63**, 275–281 (2014).
[182] Leggett A. J., *The Problems of Physics*, Oxford University Press, Oxford (1987).
[183] Lindenberg K., K. E. Shuler, V. Seshadri and B. J.West, in *Probabilistic Analysis and Related Topics, Vol. 3*, Bharucha-Reid A. T. (Ed.), Academic Press, New York (1983).
[184] Lindenberg K. and B. J. West, *The Nonequilibrium Statistical Mechanics of Open and Closed Systems*, VCH, New York (1990).
[185] Lindstedt S. L. and W. A. Calder III, "Body size and longevity in birds", *The Condor* **78**, 91–94 (1976).
[186] Lindstedt S. L. and W. A. Calder III, "Body size, physiological time, and longevity of homeothermic animals", *Quart. Rev. Biol.* **36**, 1–16 (1981).
[187] Lindstedt S. L., B. J. Miller and S. W. Buskirk, "Home range, time and body size in mammals", *Ecology* **67**, 413–418 (1986).
[188] Lindstedt S. L. and P. J. Schaeffer, "Use of allometry in predicting anatomical and physiological parameters of mammals", *Lab. Anim.* **36**, 1–19 (2002).
[189] Liu Y., J. Slotine and A. Barabási, "Controllability of complex networks", *Nature* **473**, 167–173 (2011).
[190] Lorenz E. N., "Deterministic nonperiodic flow", *J. Atmos. Sci.* **20**, 130–141 (1963).
[191] Lorenz E. N., *The Essence of Chaos*, University of Washington Press, Seattle (1993).
[192] Lowenhard P., "The mind-body problem: some neurobiological reflections", in *Reductionism and System Theory in the Life Sciences*, Hoyningen-Huene P. and F.M. Wuketits (Eds), Kluwer Academic Pub., Dordrecht (1989).
[193] Lyell C., *Principles of Geology*, Philadelphia, J. Kay (1837).
[194] Mackey M. C., *Time's Arrow*, Springer-Verlag, New York (1992).
[195] Magin R. L., *Fractional Calculus in Bioengineering*, Begell House, CN (2006).
[196] Mahmood I., "Interspecies scaling of renally secrete drugs", *Life Sci.* **63**, 2365–2371 (1998).
[197] Mainardi F., R. Gorenflo and E. Scalas, "A fractional generalization of the Poisson process", *Vietnam J. Math.* **32**, 53–64 (2004).
[198] Makarieva A. M., V. G. Gorshkov and B.-L. Li, "Revising the distributive networks models of West, Brown and Enquist (1997) and Banavar, Maritan and Rinaldo (1999): metabolic inequity of living tissue provides clues for the observed allometric scaling rules", *J. Theor. Biol.* **237**, 291–301 (2005).
[199] Makarieva A. M., V. G. Gorshkov and B.-L. Li, "Why do population density and inverse home range scale differently with body size? Implications for ecosystem stability." *Ecol. Comp.* **2**, 259–271 (2005).
[200] Manaster R. D. and S. Manaster, "Techniques for estimating allometric equations", *J. Morphol.* **147**, 299–308 (1975).
[201] Mandelbrot B. B., *Fractals, Form and Chance*, W. H. Freeman, San Francisco, CA (1977).
[202] Mandelbrot B. B., *The Fractal Geometry of Nature*, W. H. Freeman, San Francisco, CA (1982).
[203] Mandelbrot B. B., "Self-affine fractal sets", in *Fractals in Physics*, pp. 3–28, Pietronero L. and E. Tosatti (Eds), North-Holland, Amsterdam (1986).
[204] Mandelbrot B. B., *Fractals and Scaling in Finance*, Springer, New York (1997).

[205] Mantegna R. N. and H. E. Stanley, *Econophysics*, Cambridge University Press, New York (2000).

[206] Maritan A., R. Rigon, J. B. Banavar and A. Rinaldo, "Network allometry", *Geophys. Res. Lett.* **29**, 1508 (2001).

[207] Martin R. D. and P. H. Harvey, "Brain size allometry: ontogeny and phylogeny", in *Size & Scaling in Primate Biology*, Jungers W. L. (Ed.), Plenum Press, New York (1985).

[208] Mauroy B., M. Flloche, E. R. Weibel and B. Sapoval, "An optimal bronchial tree may be dangerous", *Nature* **427**, 633–638 (2004).

[209] McMahon T. A., "Size and shape in biology", *Science* **179**, 1201–1204 (1973).

[210] McMahon T. A., "Allometry and biomechanics: limb bones in adult ungulates", *Amer. Nat.* **109**, 547–563 (1975).

[211] McMahon T. A. and J. T. Bonner, *On Size and Life*, Sci. Am. Library, NY (1983).

[212] McNab B. K., *The Physiological Ecology of Vertebrates; A View from Energetics*, Comstock Publ. Ass. (2002).

[213] McNab B. K., "Ecological factors affect the level and scaling of avian BMR", *Comp. Biochem. Physiol.* **152**, 22–45 (2009).

[214] Meakin P., *Fractals, Scaling and Growth Far from Equilibrium*, Cambridge University Press, Cambridge, MA (1998).

[215] Meerschaert M. M., E. Nane and P. Vellaisamy, "The fractional Poisson process and the inverse stable subordinator", *Electron. J. Prob.* **16**, 1600–1620 (2004).

[216] Meerschaert M. M. and A. Sikorskii, *Stochastic Models for Fractional Calculus*, De Gruyter, Berlin; Boston (2012).

[217] Milgrim S., "The experience of living in cities", *Science* **167**, 1461 (1970).

[218] Miller K. S. and B. Ross, *An Introduction to the Fractional Calculus and Fractional Differential Equations*, John Wiley & Sons, New York (1993).

[219] Minamino R. and M. Takeno, "Tree branching: Leonardo da Vinci's rule versus biomechanical models", *PLOS ONE* **9**, e93535-1, -12 (2014).

[220] Mittag-Leffler G. M., "Sur la Nouvelle function", *C. R. Acad. Sci. Paris* **137**, 554 (1903) .

[221] Mohr C. O., "Comparitive populations of game, fur and other mammals", *Am. Midl. Nat.* **24**, 581–584 (1940).

[222] Montgomery D. R. and W. E. Dierich, "Channel initiation and the problem of landscape scale", *Science* **255**, 826–832 (1992).

[223] Montroll E. W. and G. Weiss, "Random walks on lattices. II", *J. Math. Phys.* **6**, 167–181 (1965).

[224] Montroll E. W. and B. J. West, "On an enriched collection of stochastic processes", in *Fluctuation Phenomena*, Montroll E. W. and J. L. Lebowitz (Eds), Studies in Statistical Mechanics, Vol. VII, North-Holland, Amsterdam (1979).

[225] Mordenti J., "Forecasing cephalosporin and monobactam antibiotic half-lives in humans from data collected in laboratory animals", *Antimicrobial Agents and Chemotherapeutics* **27**, 887–891 (1985).

[226] Morgado E., B. Günther and U. Gonzalez, "On the allometry of wings", *Revista Chilena de Hisoria Natural* **60**, 71–79 (1987).

[227] Moses M. E., S. Gorrest, A. L. Davis, A. L. Lodder and J. H. Brown, "Scaling theory for information networks", *J. Roy. Soc. Interface* **5**, 1469–1480 (2008).

[228] Murray C. D., "The physiological principle of minimum work. I. The vascular system and the cost of blood", *Proc. Nat. Acad. Sci. USA* **12**, 207–214 (1926).

[229] Murray C. D., "The physiological principle of minimum work applied to the angle of branching of arteries", *J. Gen. Physiol.* **9**, 835–841 (1926).

[230] Nelson T. R., B. J. West and A. L. Goldberger, "The fractal lung: universal and species-related scaling patterns", *Experientia* **46**, 251–254 (1990).

[231] Newman M. E. J., “The structure and function of complex networks”, *SIAM Rev.* **45**, 167 (2003).

[232] Newton I., *Mathematical Principles of Natural Philosophy* (1668); translated by A. Motte, in *Great Books of the Western World* **34**, William Benton, Encyclopedia Britannica Inc., Chicago (1952).

[233] Niklas K. J., “Plant allometry: is there a grand unifying theory?”, *Biol. Rev.* **79**, 871–889 (2004).

[234] Omori F., “On the aftershocks of earthquakes”, *J. College Sci. Imperial Univ. Tokyo* **7**, 111–200 (1894).

[235] Osorio J., M. G. Frei, D. Sornette, J. Milton and C. Lai, “Epileptic seizures; quakes of the brain?”, *Phys. Rev. E* **82**, 021919 (2010).

[236] Ott E., *Chaos in Dynamical Systems*, Cambridge University Press, New York (1993).

[237] Packard G. C., “On the use of logarithmic transformations in allometric analysis”, *J. Theor. Biol.* **257**, 515–518 (2008).

[238] Packard G. C. and T. J. Boardman, “Model selection and logarithmic transformation in allometric analysis”, *Physiol. Biochem. Zool.* **81**, 496–507 (2008).

[239] Painter P. R., “Allometric scaling of the maximum metabolic rate of mammals: oxygen transport from the lungs to the heart is a limiting step”, *Theor. Biol. Med. Model.* **2**, 31–39 (2005).

[240] Pareto V., *Cours d’Economie Politique*, Lausanne and Paris (1897).

[241] Peckham S., “New results for self-similar trees with applications to river networks”, *Water Resources Res.* **31**, 1023–1029 (1995).

[242] Peckham S. D. and V. K. Gupta, “A reformulation of Horton’s law for large river networks in terms of statistical self-similarity”, *Water Resources Res.* **35**, 2763–2777 (1999).

[243] Peng C.-K., J. Mistus, J. M. Hausdorff, S. Havlin, H. E. Stanley and A. L. Goldberger, “Long-range anti-correlations and non-Gaussian behavior of the heartbeat”, *Phys. Rev. Lett.* **70**, 1343–1346 (1993).

[244] Penrose R., *Shadows of the Mind, a Search for the Missing Science of Consciousness*, Oxford University Press, Oxford (1994).

[245] Perrin J., *ATOMS*, French edition, Livrairie Felix Alcan (1913); translated by D. Ll. Hammick, Ox Bow Press, Woodbridge, CO (1990).

[246] Peterson M., *Ergodic Theory*, Cambridge University Press, Cambridge (1983).

[247] Peters R. H., *The Ecological Implications of Body Size*, Cambridge University Press, Cambridge (1983).

[248] Piccinini N., D. Lambert, B. J. West, M. Bologna and P. Grigolini, “Nonergodic complexity management”, *Phys. Rev. E* **93**, 062301 (2016).

[249] Pilbeam D. and S. J. Gould, “Size and scaling in human evolution”, *Science* **186**, 892–901 (1974).

[250] Plotnick R. E. and J. J. Sepkoski, “A multiplicative multifractal model for orginations and extinctions”, *Paleobiol.* **27**, 126–139 (2001).

[251] Podlubny I., *Fractional Differential Equations*, Mathematics in Science and Engineering **198**, Academic Press, San Diego (1999).

[252] Poincaré H., *Mémoire sur les courves d éfinies parles equations différentielles, I–IV, Oevre 1*, Gauthier–Villars, Paris (1888).

[253] Pramukkul P., A. Svenkeson, P. Grigolini, M. Bologna, B. J. West, “Complexity and the fractional calculus”, *Advances in Mathematical Physics* **2013**, 498789 (2013).

[254] Preston F. W., “The canonical distribution of commonness and rarity”, *Ecology* **43**, 185–215 (1962).

[255] Prothero J. W., “Scaling of cortical neuron density and white matter volume in mammals”, *J. Brain Res.* **38**, 513–524 (1997).

[256] Prigogine I., *Thermodynamics of Irreversibile Processes*, 2nd revised edition, Wiley Interscience, New York (1961); 1st edition (1955).
[257] Prigogine I., "Time, structure, and fluctuations", *Science* **201**, 777–785 (1978).
[258] Prigogine I. and I. Stengers, *Order Out of Chaos: Man's New Dialogue with Nature*, Bantam Books, Toronto (1984).
[259] Que C. L., C. M. Kenyon, R. Olivenstein, P. T. Macklem and G. N. Maksym, "Homeokinesis and short-term variability of human airway caliber", *J. Appl. Physiol.* **91**, 1131–1141 (2001).
[260] Rabotnov Yu. N., *Elements of Hereditary Solid Mechanics*, MIR Pub., Moscow (1977).
[261] Raia P. and M. Fortelius, "Cope's law of the unspecified, Cope's rule, and weak directionality in evolution", *Evol. Eco. Res.* **15**, 1–0 (2013).
[262] Rashevsky N., *Mathatical Biophysics Physico-Mathematical Foundations of Biology, Volume Two*, 3rd edition, Dover Publications, New York (1960).
[263] Raup D. M. and J. J. Skepkoski Jr., "Mass extinction in the Marine Fossil record", *Science* **215**, 1501–1503 (1982).
[264] Rayner J. M. V., "A new approach to animal flight mecahnisms", *J. Exp. Bio.* **80**, 17–54 (1979).
[265] Reich P. B., M. G. Tjoelker, J. Marchado and J. Oleksyn, "Universal scaling of respiratory metabolism, size and nitrogen in plants", *Nature* **439**, 457–461 (2006).
[266] Reichl L. E., *A Modern Course in Statistical Physics*, John Wiley & Sons, New York (1998).
[267] Reiss M. J., *The Allometry of Growth and Reproduction*, Cambridge University Press, Cambridge (1989).
[268] Richardson L. F., "Atmospheric diffusion shown on a distance-neighbour graph", *Proc. Roy. Soc. London A* **110**, 709 (1926).
[269] Richter J. P. (Ed.), *The Notebooks of Leonardo da Vinci, Vol. 1*, Dover, New York (1970); unabridged edition of the work first published in London in 1883.
[270] Rigon R., I. Rodriguez-Iturbe and A. Rinaldo, "Feasible optimality implies Hack's law", *Water Resources Res.* **32**, 3367–3374 (1998).
[271] Rikvold P. A. and R. K. P. Zia, "Punctuated equilibria and $1/f$ noise in a biological coevolution model with individual-based dynamics", *Phys. Rev. E* **68**, 031913 (2003).
[272] Rinaldo A., J. R. Banavar and A. Maritan, "Trees, networks and hydrology", *Water Resources Res.* **42**, 1–19 (2006).
[273] Robersts D. C. and D. L. Turcotte, "Fractality and self-organized criticality of wars", *Fractals* **6**, 351 (1998).
[274] Rocco A. and B. J. West, "Fractional calculus and the evolution of fractal phenomena", *Physica A* **265**, 535 (1999).
[275] Rodriguez-Iturbe I. and A. Rinaldo, *Fractal River Basins. Chance and Self-organization*, Cambridge University Press, Cambridge, UK (1997).
[276] Samorodnitsky G. and M. S. Taqqu, *Stable Non-Gaussian Random Processes*, Chapman & Hall, New York (1994).
[277] Sacher G. A., "Relation of lifespan to brain weight and body weight in mammals", in *Ciba Foundation Colloquium on Aging, Vol. 1*, Wolstenholme G. E. W. (Ed.) (1959).
[278] Sagar B. S. and T. L. Tien, "Allometric power-law relationships in a Hortonian fractal digital elevation model", *Geophys. Res. Lett.* **31**, L06501, 1–4 (2004).
[279] Saichev A. I. and G. M. Zaslavsky, "Fractional kinetic equations: solutions and applications", *CHAOS* **7**, 753–764 (1997).
[280] Samko S. G., A. A. Kilbas and O. I. Marichev, *Fractional Integrals and Derivatives*, Gordon and Breach Science Pub., USA (1993).
[281] Sapoval B., *Universalitiés et Fractales*, Flammarion, Paris (1997).
[282] Savage V. M., J. P. Gillooly, W. H. Woodruff, G. B. West, A. P. Allen, B. J. Enquist and J. H. Brown, "The predominance of quarter-power scaling biology", *Func. Ecol.* **18**, 257–282 (2004).

[283] Savage V. M., E. J. Deeds and W. Fontana, "Sizing up allometric scaling", *PLoS Compt. Biol.* **4** (9), e1000171 (2008).

[284] Sawada Y., M. Hanano, Y. Sugiyama and T. Iga, "Prediction of disposition of beta-lactam antibiotics in humans from pharmacokinetic parameters in animals", *J. Pharmacokin Biopharm.* **12**, 241–261 (1984).

[285] Scheidegger A. E., "On the evolution of river nets", *Bull. Int. Assoc. Sci. Hyd.* **XV**, 1 (1970).

[286] Schenzle A. and H. Brand, "Multiplicative stochastic processes in statistical physics", *Phys. Rev. E* **20**, 1628–1647 (1979).

[287] Schlenska G., "Volumen und Oberflachenmessungen an Gehiren verschiedener Saugetiere im Vergleich su einem errechneien Modell", *J. Hirnforsch* **15**, 401–408 (1974).

[288] Schmidt-Nielsen K., *Scaling, Why Is Animal Size so Important?*, Cambridge University Press, Cambridge (1984).

[289] Schmidt-Nielsen K., *Animal Physiology*, Cambridge University Press, Cambridge (1997).

[290] Schrödinger E., *What Is Life?*, Cambridge University Press, New York (1995); based on lectures delivered under the auspices of the Dublin Institute for Advanced Studies at Trinity College, Dublin, in February (1943).

[291] Schweber S. S., "Physics, community and the crisis in physical theory", *Physics Today* **46** (11), 34 (November 1993).

[292] Scott Blair S. G., B. C. Veinoglou and J. E. Caffyn, "Limitations of the ... theory of quasi-properties", *Proc. Roy. Soc. Ser. A* **187**, 69 (1947).

[293] Seshadri V. and B. J. West, "Fractal dimensionality of Lévy processes", *Proc. Nat. Acad. Sci. USA* **79**, 4501–4505 (1982).

[294] Shannon C. E., "A mathematical theory of communication", *The Bell System Journal* **XXVII**, 379–423 and 623-56 (1948).

[295] Shea B., "Relative growth of the limbs and trunk in the African apes", *Am. J. Phys. Anthopol.* **56**, 179–202 (1981).

[296] Shlesinger M. F., B. J. West and J. Klafter, "Lévy dynamics of enhanced diffusion: application to turbulence", *Phys. Rev. Lett.* **58**, 1100 (1987).

[297] Shlesinger M. F. and B. J. West, "Complex fractal dimension of the bronchial tree", *Phys. Rev. Lett.* **67**, 2106–2108 (1991).

[298] Silva da J. K. L., G. J. M. Garcia and L. A. Barbosa, "Allometric scaling laws of metabolism", *Phys. Life Reviews* **3**, 229–261 (2006).

[299] Smith R. J., "Rethinking allometry", *J. Theor. Biol* . **87**, 97–111 (1980).

[300] Smith R. J., "Logarithmic transformation bias in allometry", *Am. J. Phys. Anthropol.* **90**, 215–228 (1993).

[301] Smoluchowski M., "Zur kinetischen Theorie der Brownschen Molekularbewegung und der Suspensionen", *Ann. Physik* **21**, 756 (1906).

[302] Snell O., "Die Abhängigkeit des Hirngewichts von dem Körpergewicht und den geistigen Fähigkeiten", *Arch. Psychiatr.* **23**, 436–446 (1892).

[303] Snell K. (Ed.), *Understanding the Control of Metabolism*, Portland, London (1997).

[304] Sneppe K., P. Bak, H. Flybjerg and M. H. Jensen, "Evolution as a self-organized critical phenomenon", *Proc. Nat. Acad. Sci. USA* **92**, 5209–5213 (1995).

[305] Sokolov I. M., "Linear response to perturbation of nonexponential renewal process: a generalized master equation approach", *Phys. Rev. E* **73**, 067102 (2006).

[306] Sokolov I. M., J. Klafter and A. Blumen, "Fractional kinetics", *Physics Today*, 48–54 (November 2002).

[307] Sokolov I. M. and J. Klafter, "From diffusion to anomalous diffusion: a century after Einstein's Brownian motion", *Chaos* **15**, 026103 (2005).

[308] Solé R. V., S. C. Manrubia, M. Benton and P. Bak, "Self-similarity of extinction statistics in the fossil record", *Nature* **388**, 764–767 (1997).
[309] Sornette D., "Discrete scale invariance and complex dimensions", *Phys. Rept.* **297**, 239 (1998).
[310] Speakman J. R., "Body size, energy metabolism and lifespan", *J. Exp. Bio.* **208**, 1717–1730 (2005).
[311] Svenkeson A., M. T. Beig, M. Turalska, B. J. West and P. Grigolini, "Fractional trajectories: decorrelation versus friction", *Physica A* **392**, 5663 (2013).
[312] Taskinen S. and D. I. Warton, "Robust estimation and inference for bivariate line-fitting in allometry", *Biometrical Journal* **53**, 652–677 (2011).
[313] Taylor L. R., "Aggregation, variance and the mean", *Nature* **189**, 732–735 (1961).
[314] Taylor L. R. and R. A. J. Taylor, "Aggregation, migration and population mechanics", *Nature* **265**, 415–421 (1977).
[315] Taylor C. R. and I. P. Woiwod, "Temporal stability as a density-dependent species characterization", *J. Animal Ecol.* **49**, 209–224 (1980).
[316] Taylor C. R. and E. R. Weibel, "Design of the mammalian respiratory system, 1. Problem and strategy", *Respir. Physiol.* **44**, 1–10 (1981).
[317] Thompson D. W., *On Growth and Form*, The complete revised edition; originally published by Cambridge University Press, UK (1942); Dover Pub., New York (1992).
[318] Tippett M. K. and J. E. Cohen, "Tornado outbreak variability follows Taylor's power law of fluctuation scaling and increases dramatically with severity", *Nature Com.* **7**, 10668 (2016).
[319] Tower D. B., "Structural and functional organization of mammalian cerebral cortex the correlation of neurone density with brain size", *J. Comp. Neurol.* **101**, 9–52 (1954).
[320] Turalska M., M. Lukovic, B. J. West and P. Grigolini, "Complexity and synchronization", *Phys. Rev. E* **80**, 021110-1 (2009).
[321] Turalska M., P. Grigolini and B. J. West, "Role of committed minorities in times of crisis", *Sci. Rept.* **3**, 1–8 (2013).
[322] Turalska M. and B. J. West, "Fractional dynamics of complex networks", under review.
[323] Turcotte D. L., *Fractals and Chaos in Geology and Geophysics*, Cambridge University Press, Cambridge (1992).
[324] Vlad M. O., F. Moran, V. T. Popa, S. E. Szedlacsek and J. Ross, "Functional, fractal nonlinear response with application to rate presses with memory, allometry, and population genetics", *PNAS* **104**, 4798–4803 (2007).
[325] Von Bertalanffy L., "Quantitative laws in metabolism and growth", *Q. Rev. Biol.* **32**, 217–231 (1957).
[326] Wang M. C. and G. E. Uhlenbeck, "On the theory of the Brownian motion II", *Rev. Mod. Phys.* **17**, 323 (1945).
[327] Warton D. I., I. J. Wright, D. S. Falster and M. Westoby, "Bivariate line fitting methods for allometry", *Biol. Rev.* **85**, 259–291 (2006).
[328] Watts D. J., *Small Worlds*, Princeton University Press, Princeton, NJ (1999).
[329] Weaver W., "Science and complexity", *American Scientist* **36**, 536–544 (1948).
[330] Weibel E. R. and D. M. Gomez, "Architecture of the human lung", *Science* **137**, 577–585 (1962).
[331] Weibel E. R., *Symmorphosis: On Form and Function in Shaping Life*, Harvard University Press, Cambridge, MA (2000).
[332] Weibel E. R., "The pitfalls of power laws", *Nature* **417**, 131–132 (2002).
[333] Weinberg S., *Dreams of a Final Theory*, Pantheon Books, New York (1992).
[334] Weiss G. H., *Aspects and Applications of the Random Walk*, North-Holland, Amsterdam (1994).
[335] Werner G., "Fractals in the nervous system: conceptual implications for theoretical neuroscience", *Front. Physiol.* **1**, 15 (2010). doi: 10.3389/fphys.2010.00015.

[336] West B. J. and V. Seshadri, "Linear systems with Lévy fluctuations", *Physica A* **113**, 293–216 (1982).
[337] West B. J., *The Importance of Being Nonlinear*, Springer, Berlin (1985).
[338] West B. J., V. Barghava and A. L. Goldberger, "Beyond the principle of similitude: renormalization in the bronchial tree", *J. Appl. Physiol.* **60**, 1089–1097 (1986).
[339] West B. J. and A. L. Goldberger, "Physiology in fractal dimensions", *Am. Sci.* **75**, 354–365 (1987).
[340] West B. J., *Fractal Physiology and Chaos in Medicine*, Studies of Nonlinear Phenomena in Life Science, Vol. 1, World Scientific, Singapore (1990).
[341] West B. J., "Physiology in fractal dimension: error tolerance", *Ann. Biomed. Eng.* **18**, 135–149 (1990).
[342] West B. J. and M. Shlesinger, "The noise in natural phenomena", *Amer. Sci.* **78**, 40–45 (1990).
[343] West B. J., *The Lure of Modern Science: Fractal Thinking*, Studies of Nonlinear Phenomena in Life Science, Vol. 3, World Scientific, Singapore (1999).
[344] West B. J., *Physiology, Promisucity and Prophecy at the Millennium: A Tale of Tails*, Studies of Nonlinear Phenomena in Life Science, Vol. 7, World Scientific, Singapore (1999).
[345] West B. J., M. Bologna and P. Grigolini, *Physics of Fractal Operators*, Springer, Berlin (2003).
[346] West B. J., *Where Medicine Went Wrong*, Studies of Nonlinear Phenomena in Life Science, Vol. 11, World Scientific, Singapore (2006).
[347] West B. J., E. L. Geneston and P. Grigolini, "Maximizing information exchange between complex networks", *Phys. Rept.* **468**, 1–99 (2008).
[348] West B. J., "Fractal physiology and the fractional calculus: a perspective", *Front. Physiol.* **1**, 12 (2010). doi: 10.3389/fphys.2010.00012.
[349] West B. J. and D. West, "Origin of allometry hypothesis", in *Applications of Fractional Calculus*, Metzler R. (Ed.) (2011).
[350] West D. and B. J. West, "Are allometry and macroevolution related?", *Physica A* **390**, 1733–1736 (2011).
[351] West D. and B. J. West, "Stochastic origin of allometry", *EPL* **94**, 38005 (2011).
[352] West B. J. and P. Grigolini, *Complex Webs: Anticipating the Improbable*, Cambridge University Press, Cambridge, UK (2011).
[353] West B. J. and D. West, "Fractional dynamics of allometry", *Fractional Calculus & Applied Analysis* **15**, 127 (2012).
[354] West B. J. and D. West, "Stochastic ontogenetic growth model", *EPL* **97**, 48002 (2012).
[355] West D. and B. J. West, "On allometry relations", *Int. J. Mod. Phys. B* **26**, 1230013-1 (2012).
[356] West D. and B. J. West, "Physiological time: a hypothesis", *Phys Life Revs.* **10**, 210–224 (2013).
[357] West B. J., *Fractal Physiology and Chaos in Medicine*, World Scientific (2013).
[358] West B. J., M. Turalska and P. Grigolini, *Network of Echoes; Imitation, Innovation and Invisible Leaders*, Springer, Berlin (2014).
[359] West B. J., "Fractional calculus view of complexity: a tutorial", *Rev. Mod. Phys.* **86**,1169–1184 (2014).
[360] West B. J., "A mathematics for medicine: the network effect", *Front. Fractal Phys.* **5**, 1–17 (2014).
[361] West B. J., M. Turalska and P. Grigolini, "Fractional calculus ties the microscopic and macroscopic scales of complex network dynamics", *N. J. Phys.* **17**, 045009 (2015).
[362] West B. J., *Fractional Calculus View of Complexity, Tomorrow's Science*, CRC, Boca Raton (2016).
[363] West B. J., "Information forces", *J. Theor. & Comp. Sci.* **3**, 144 (2016).
[364] West B. J., *Simplifying Complexity; Life is Uncertain, Unequal and Unfair*, Bentham Books, UAE (2016).

[365] West G. B., Brown J. H. and Enquist B. J., "A general model for the origin of allometric scaling laws in biology", *Science* **276**, 122–124 (1997).
[366] West G. B., "The origin of universal scaling laws in biology", *Physica A* **263**, 104–113 (1999).
[367] West G. B., J. H. Brown and B. J. Enquist, "The fourth dimension of life: fractal geometry and allometric scaling of organisms", *Science* **284**, 1677 (1999).
[368] West G. B., *et al.*, "The origin of universal scaling laws in biology", in *Scaling in Biology*, pp. 87–112, Brown J. H. and G. B. West (Eds), Oxford University Press, Oxford (2000).
[369] West G. B., J. H. Brown and B. J. Enquist, "A general model for ontogenetic growth", *Nature* **413**, 628–631 (2001).
[370] West G. B., V. M. Savage, J. Gillooly, B. J. Enquist, W. H. Woodruff and J. H. Brown, "Why does metabolic rate scale with body size?", *Nature* **421**, 712 (2003).
[371] West G. B. and J. H. Brown, "Life's universal scaling laws", *Physics Today*, 36–42 (September 2004).
[372] White J. F. and S. J. Gould, "Interpretation of the coefficient in the allometric equation", *The Am. Natur.* **99**, 5–18 (1965).
[373] White C. R. and R. S. Seymour, "Allometric scaling of mammalian metabolism", *J. Exp. Biol.* **208**, 1611–1619 (2005).
[374] White C. R., P. Cassey and T. M. Blackburn, "Metabolic allometry exponents are not universal", *Ecology* **88**, 315–323 (2007).
[375] Whiteside D. T., "The mathematical principles underlying Newton's *Principia Mathematica*", *Jour. for the History of Astronomy* **1**, 116–138 (1970).
[376] Wiener N., "Time, communication, and the nervous system", *Ann. NY Acad. Sci.* **50**, 197 (1948).
[377] Wiener N., *Cybernetics*, MIT Press, Cambridge, MA (1948).
[378] Williams C. B., *Patterns in the Balance of Nature and Related Problems in Quantitative Ecology*, Academic Press, New York (1964).
[379] Willis J. C., *Age and Area*, Cambridge University Press, Cambridge (1922).
[380] Wilson K. G., "The renormalization group: critical phenomena and the Kondo problem", *Rev. Mod. Phys.* **47**, 773–840 (1975).
[381] Wilson T. A., "Design of the bronchial tree", *Nature*, 668 (1967).
[382] Wilson K. G. and J. Kogut, "The renormalization group and the $\epsilon$ expansion", Phys. Rep. **12**, 75 (1974).
[383] Winfree A. T., *Timing of Biological Clocks*, Scientific American Library (1987).
[384] Woodward G., B. Ebenman, M. Emmerson, J. M. Montoya, J. M. Olesen, A. Valido and P. H. Warren, "Body size in ecological networks", *TRENDS in Ecol. and Evol.* **20**, 402–409 (2005).
[385] Yates C. A., R. Erbana, C. Escudero, I. D. Couzin, J. Buhl, I. G. Kevrekidis, P. K. Mainia and D. J. T. Sumpter, "Inherent noise can facilitate coherence in collective swarm motion", *PNAS USA* **106** (14), 5464–5469 (2009).
[386] Zar J. H., "Calculation and miscalculation of the allometric equation as a model in biological data", *BioScience* **18**, 1118–1120 (1968).
[387] Zaslavsky G. M., "Renormalization group theory of anomalous transport in systems with Hamiltonian chaos", *CHAOS* **4**, 25–33 (1994).
[388] Zaslavsky G. M., "Chaos, fractional kinetics, and anomalous transport", *Phys. Rept.* **371**, 461–580 (2002).
[389] Zernike J., *Entropy, the Devil on the Pillow*, Kluwer–Deventer, Amsterdam (1972).
[390] Zhang K. and J. Sejnowski, "A universal scaling law between gray matter and white matter of cerebral cortex", *PNAS* **97**, 5621–5626 (2000).
[391] Zolotarev V. M. *One-Dimensional Stable Distributions*, Trans. Math. Mono. **65**, Am. Math. Soc., Providence, RI (1986).

# Index

Bei Fragen zur Produktsicherheit wenden Sie sich bitte an:
If you have any questions regarding product safety, please contact:

Walter de Gruyter GmbH
Genthiner Straße 13
10785 Berlin
productsafety@degruyterbrill.com